AF400892

STURMTIGER

THE COMBAT HISTORY OF STURMMÖRSER KOMPANIES 1000-1002

LEE ARCHER · TIMM HAASLER

Published in 2021 by Panzerwrecks Limited

Design by Lee Archer
Artwork by Felipe Rodna
Map by Simon Vosters
Printed by Finidr. s.r.o.
Website www.panzerwrecks.com

Panzerwrecks Limited
Great Priors
Church Street
Old Heathfield
Sussex TN21 9AJ
United Kingdom
lee@panzerwrecks.com

Contents

Acknowledgements

We would like to thank the following individuals and establishments for their assistance in preparing this book:

Robert Edwards, Javier Tome, Darren Neely, Felipe Rodna, Simon Vosters and Hilary Doyle.

This publication would not have been possible without information and photographs from the following people:

US National Archives (NARA), US Army Heritage & Education Center (USAHEC), Library and Archives Canada (LAC), The Tank Museum (TTM), National WW2 Museum (NWW2M), North Carolina Museum of History (NCMH), Bayerische Staatsbibliothek München (BSB), Russian Central Military Archives (TsAMO), Archive of Modern Conflict (AMC), Dan Ballou, Viktor Kulikov, Vyacheslav Kozitsyn, Charles Lemons, Peter Kocsis, Jonathan Holt, Richard Hunter, Gail Boyd, Jim Norian, Paul Johnson, Tom Ligon, Karlheinz Münch, Darren Neely, Reimer Röbel, Jeff Scott, David Brown, Richard Hedrick, Warren Watson Oliver Lörscher and Wolfgang Grote and Josef Wunderlich.

Introduction

The Sturm-Mörser was an impressive weapon but was another waste of valuable resources. With only eighteen manufactured, there were never enough for large-scale use. Instead, they were used by three independent companies, which were not viable from a supply and repair perspective. After their service in Warsaw, the weapon's shortcomings were documented, such as the slow rate of fire and the vehicle's weight. After Warsaw, the Sturm-Mörser did not see combat in the urban environment for which it was designed.

The effect of its high-explosive projectile was devastating and could flatten a building or knock tanks over like toys. However, why was it necessary to give the Sturm-Mörser such thick armour if it was not used at short range? Furthermore, the Tiger chassis was overburdened, slow and liable to break down.

The change of branch from Panzer to artillery shows that the Panzertruppen could not do anything with the vehicle. And Hitler reserved the right to limit the use of the vehicle. It might have made more sense to have integrated the Sturm-Mörser into Tiger I units, where they had the relevant repair and recovery equipment and training, and to a lesser extent, the spare parts.

Several books have been published about the Sturm-Mörser, mostly monographs containing the same pictures, usually in poor quality. Little is said about the units that used them. This book strives to document this in as much detail as preserved records and photographs allow.

We have been able to document nine of the eighteen vehicles. Felipe Rodna has created diptych artworks of each of these, working closely with us to attain the most authentic artworks of the Sturm-Mörser to date. In addition, Simon Vosters has created a separate combat map.

Timm and I hope you enjoy this book.

Lee Archer, Old Heathfield, July 2021

Designation

Sturm-Tiger mit 38cm Raketenwerfer 61 (RW 61)
Sturm-Mörser
Tiger-Mörser, Panzermörser mit 38cm RW 61 auf Tiger I
Sturmmörserwagen 606/4.[1]

Development

The starting point for the development of the Sturm-Mörser was the Raketen-Tauchgranate (RTg) 38cm developed by Rheinmetall-Borsig for the Kriegsmarine. The projectiles were fired from a tube mounted on a rotating platform. In order to use the weapon and make it more mobile, the Kriegsmarine planned to mount it on a self-propelled platform; the army also took an interest in this idea. For land warfare, i.e. to *"engage fortified positions,"* the range of the projectile was increased from 3000 to 5000 metres and employed an impact fuse.

When repurposed to this end, the weapon was designated 38cm Raketenwerfer 61 (38cm Rocket Launcher 61). A suitable chassis was sought for the weapon, capable of cross-country movement and the ability to carry heavy loads (due, in part, to the weight of the ammunition). Because of these demands, the only candidate at the time was the Tiger I chassis. In an effort to avoid straining the Tiger's production program, only vehicles which had been returned from the front for repair and overhaul were to be converted, which could be retrofitted at a relatively low cost.[2]

Following meetings with Hitler on 4 and 5 August 1943, Albert Speer noted that he considered the proposal to employ a Tiger chassis using the naval 38cm projector, extremely valuable. In addition, Hitler agreed with General Guderian's suggestion to produce a test vehicle. As a planning target, ten vehicles were to be produced monthly.[3]

The mild-steel prototype[4] was presented to Hitler at the Arys Training Ground in East Prussia on 20 October 1943.[5] Hitler was impressed by the design, but still could not decide whether to approve it for production. It was to be deployed in combat first.

The Army General Staff was not particularly enthusiastic about the Sturm-Mörser, and would rather have seen an increase in the production of the regular Tiger with the proven 8·8cm main gun. Therefore, it is not surprising that in its 'Study on Armour 1944,' the General Staff recommended various vehicle quotas be reduced.[6] In the column listing 'on-hand equipment,' the 'Tiger projector 38cm' was annotated with *"1 item completed,"* but in the column, 'future equipment,' nothing was entered. Apparently, the General Staff believed it could do without this weapon. However, the proposal by the high command did not sit well with Hitler.[7]

In meetings on 6 and 7 April 1944, Hitler directed that the next demonstration of the Tiger projector be accelerated. This occurred on 20 April in the form of a drive-by on the Reichsautobahn near Klessheim Castle. The demonstration impressed Hitler in such a way that he ordered the construction of twelve launchers and superstructures using overhauled Tiger I chassis, and a basic load of ammunition was to be manufactured.[8] Although there was still resistance against the 38cm Sturm-Mörser within the high command, it was Hitler who ultimately decided what was built.[9]

The development of the Raketenwerfer 61 took at the Rheinmetall-Borsig Sömmerda plant; Brandenburger Eisenwerke manufactured the armour at Kirchmöser (Brandenburg); and the vehicle was assembled at Alkett (Spandau/Berlin).[10, 11]

Two production vehicles were finished at Alkett in August 1944. Both vehicles and the prototype were accepted by the Army weapons inspections office and sent directly to Warsaw, where they were assigned to Sturm-Mörser-Kompanie 1000. All subsequent vehicles were also accepted by the weapons inspection office, but they were then turned over to the Inspector General of the Armoured Forces for issue.

In the middle of August, Albert Speer reported to Hitler that seven more Sturm-Mörsers were to be produced by Alkett in Berlin between 15 and 21 September 1944. Hitler attached great importance to the weapon for its use in a special-purpose role and initially considered a minimum production of 300 projectiles a month as necessary. After completion of its mission in the East, the prototype was to be fitted with an armoured superstructure at Alkett before being transferred to the West.[12]

On 23 September 1944, Speer reported to Hitler that a total of 10 Sturm-Mörsers had been converted at Alkett during September under the leadership of Obermeister Hahne. Hitler immediately ordered the further conversion of five Sturm-Mörsers vehicles a month.[13] Despite that, no more vehicles were converted until December 1944, when five were finished at Alkett. Those five vehicles were not assigned to combat units in 1944.

A report dated 1 January 1945, shows that the 13 previously converted Sturm-Mörsers were assigned to Sturm-Mörser-Kompanie 1000, 1001 and 1002 (four each) and Panzer-Ersatz- und Ausbildungs-Abteilung 500 (prototype).[14]

On 5 January 1945, Hitler made the decision to continue the 'Tiger Mörser' on repaired 'Tiger I' chassis until a lighter self-propelled gun using the same weapon was designed.[15]

Above: The prototype Sturm-Mörser was assembled on a Tiger Ausf.E chassis with 'Feifel' air-cleaners and rubber-tyred roadwheels. It is shown here with the outer roadwheels removed and transport tracks fitted. *K.Münch*

Production and Allocation Overview

Month	Alkett	Wp ins of	In 6			Sturm-Mörser Kompanie1000		Sturm-Mörser Kompanie1001		Sturm-Mörser Kompanie1002		Pz-Ers. Abt.500		Wp Ins Borsig	
		Acceptance	+	-	=	=	Chassis	=	Chassis	=	Chassis	=	Chassis	=	Chassis
October 1943	1														
August 1944	2	3				3	Prototype: 250091: 250230	0		0		0		0	
September 1944	10	10				3	Prototype: 250091: 250230	0		0		0		0	
October 1944			10	8	2	4	250091: 250230: 250237: 251168	4		0		2	Prototype: 251174	1	250237
November 1944				1	1	3	?: 250237: 251168	4		0			Prototype: 251174	1	250237
December 1944	5	5	1	2		4	250091: 250230: 250237: 251168	4		4	250237: 251174: 25xxxx: 25xxxx	1	Prototype	0	
January 1945			5		5	4		4		4		1			
February 1945				2	3	4		4		6		1			
March 1945				2	1	5	Oberembt (250237)	4		6		1			
April 1945							Drolshagen	3	Rogäsen	3	Ebendorf (251174): Menden: Brumby				
Total	18	18	16	15	12										

Remarks:
* = either 250091 or 250230 was returned to Germany and assigned to In 6 in December.

Sources:
BA-MA RH 10/349, page 316; BA-MA RH 10/350, page 156.
NARA T78, R169, Frame 6106514 and 6106698.
https://forum.valka.cz/topic/view/6453/Sturmmorser-Tiger
No author: Waffen Revue, Vol. 35, page 5540.

Hitler and top brass inspect new weapons at Arys, East Prussia, on 20 October 1943. The Sturm-Mörser makes for an imposing sight, towering above them. The 'L'-shaped brackets on the glacis plate were presumably to fix a canvas cover over the weapon. ***BSB München***

The thickness and angles of the armour plates were painted onto the new superstructure, but not the Tiger chassis. The 'Feifel' air cleaners and their associated ducting are visible on the engine deck and rear wall. ***BSB München***

Sturm-Mörser Prototype
Arys Training Ground, East Prussia
20 October 1943

Hitler looks into the large loading hatch while Dipl.-Ing. Paul Michaels, Oberingenieur and Chefkonstrukteur of Alkett explains the details. The large slab of armour plate bolted to the bow armour is a feature of some Sturm-Mörsers. The thickness and angle of the glacis armour are painted on: 150mm at 45°. Note that the driver has only one periscope fitted; the other has a steel mesh instead. The two periscopes were probably set at different angles in the same manner as the Jagdpanzer 38. *BSB München*

Technical data

Length of hull: 6·31m[16]
Width: 3·57m
Height: 2·85m
Weight: 65 tons
Hull: Welded with 150mm front, 80mm side and 80mm rear armour.[17]
Ground clearance: 0·47m
Ground pressure: 15 N/cm^2
Drive train: Interleaved running gear with four inner and four outer pairs of wheels per side, suspended from torsion bars via swing arms. On the first and last swingarm, one-sided shock absorbers were mounted in the hull. Track tensioning was carried out mechanically via the idler wheel. For rail transport, narrower transport track had to be fitted. These tracks were provided by the German railways for Tiger transport railway cars.[18]
Climbing ability: 0·79m
Gradient: 35°
Ditching capability: 2·50m
Fording capability: 1·20m
Engine/Transmission: Rear-mounted Maybach HL 230 P 45 - 12-cylinder V 600 four stroke petrol engine that delivered power via a universal joint and a wet multi-disc clutch to the Maybach 'Olvar' 8 forward and 4 reverse gearbox fitted in the bow.
Engine output: 522 kW (700 HP)
Capacity: 23 litres
Maximum speed: 45 km/h[19]
Fuel capacity: 540 litres
Range on roads: 100 km[20]
Range cross-country: 60 km
Armament: 38cm Raketenwerfer 61 L 5·4[21]
Traverse and elevation: 10° left and right. Elevation: 0° to +85° [22]
Ammunition: 14 projectiles[23]
Secondary Armament: 7·92mm MG 34
Traverse and elevation: 15° left and right. Elevation: -7° to +20°
Ammunition: 600 rounds
Radio: FuG 5[24]

Fighting compartment

The fighting compartment was a welded box-shaped construction. Despite its large size, there was only limited storage space for the single-piece 38cm projectiles, which were 1·49m long. Stowage racks were provided for six rounds each on the sides of the fighting compartment. The projectiles were mounted in pairs, one pair on top of the other. To remove the 345kg rounds from the racks, a simple gantry system was used, employing rails fitted to the roof. To line up the projectile for loading, a tray with rollers was used. In addition to the 12 stowed rounds, one could be pre-loaded in the barrel and another placed on the loading tray, resulting in a maximum basic load of 14 rounds.

A rotating crane was used to load ammunition. Mounted on the rear right of the fighting compartment, it could lift and rotate the rounds from the ground to the loading hatches in the roof. There were two hatches, with the smaller one being opened first towards the rear. Then, using the crane, the crew moved the large loading hatch to the side. When loading, the ammunition was placed on the right side of the vehicle, and the projectiles individually raised. Once at the correct height, the crane was swung over the loading bay; the ammunition was lowered into the fighting compartment to the loading tray level. In addition to two loading hatches in the roof, there was a ventilator opening to the front right and an additional roof hatch to the left rear with a rotating periscope and openings for a scissors periscope. Within the small loading hatch, was a mount for a close-in defensive weapon - 'Nahverteidigungswaffe' - which could be operated from inside the vehicle. Pistol ports were located towards the front of the fighting compartment on both sides. An outward opening hatch was fitted at the rear of the fighting compartment.

The 38cm Raketenwerfer 61 was located right of the centre of the sloped front of the compartment in a spherical mount. To its right was an MG 34 in a ball mount. To the left of the main weapon were two periscopes for the driver, set at different angles for the best field of view. Above this was an opening for the gunsight.

The fighting compartment was bolted to the hull with a series of brackets. The bolt heads were visible from the outside, but could only be loosened from the inside. Two additional sets of mounting bolts can be seen at the very front of the fighting compartment, where it joins the hull. After removing the bolts, the fighting compartment could be lifted from the hull using four lifting hooks. The hull and running gear was essentially that of the final version of the Tiger Ausf.E. In comparison to the regular Tiger, mobility was seriously reduced due to the increase in weight. [25]

38cm Raketenwerfer 61

Armament: 38cm Raketenwerfer 61
Ammunition: 38cm Raketen Sprenggranate[26]
Length: 1507mm[27]
Weight: 345kg[28]
Explosive substance: Approx. 125kg TNT
Propellant charge: Approx. 40kg of diglycol powder in rod form
Maximum range: 5500m [29, 30]
Sight: PaK ZF 3 X 8

The projectile consisted of two parts: [31] A seamless, longitudinally welded shell casing and a ballistic hood. The ballistic hood was connected to the combustion chamber with a clamping ring. The explosive weight was 135 kiloponds (kp)[32]. At the front of the projectile was the point-detonating (percussion) fuse. The combustion chamber was filled with about 40 kp diglycol powder in rod form.[33] This charge was fired from the front of the combustion chamber with the help of coarse-grained black powder. An igniter was used at the bottom of the projectile.

To load, the barrel was depressed to 0°. The projectile was placed on a loading tray in front of the open breech. With the aid of a two-handled ram, the crew pushed the projectile into the barrel until the leading guide ring engaged the grooves (nine in all), which were slightly wider at the base of the tube to allow for an easier fit. A spring-loaded plunger was actuated at

Above: The igniter. Note the splines at the base of the projectile. *NARA*

that point, preventing the round from sliding back. The ignition cartridge was then inserted into a holder on the breech plate. With a hand crank attached to the bottom of the breech plate, a rack-engaging pinion was engaged to close it horizontally from right to left.

Once the projectile was loaded, aiming could proceed. The elevation wheel was located to the left of the bottom of the weapon; the elevation range was from 0° to +85°. The vehicle was first positioned in the general direction, followed by final adjustments by means of a traverse handwheel, located above the breech. Traverse range was 10° either left and right. If required, the commander was able to determine the target range with an external optical rangefinder. Even with a well-trained crew, it was difficult to fire in less than 10 minutes.

Firing was accomplished with an ignition cartridge, whose burning jet ignited the ignition charge. The rocket left the barrel at a speed of about

300m/sec. The gases emerging from the 32 combustion chamber nozzles at the rear of the projectile developed a pressure of about 30 kp/cm^2. A simple solution was found to reduce the recoil and not restrict the already limited space in the fighting compartment. The gun barrel consisted of an (inner) core liner and an outer tube shell. The inner barrel and jacket were held at the back by holding segments and at the muzzle by semi-circular rings. Holes were drilled in the rings: 15 in the upper segment and 16 in the lower.[34] The space between the barrel and the gun jacket absorbed the propellant gases of the igniting projectile and the pressure could largely escape to the front. As a result, the chassis and the gun mount bore only a small part of the recoil energy.

In order to ease the burden on the gunner when elevating the weapon, a counterweight was welded to the muzzle of some vehicles.[35]

Ammunition

The Raketen-Spreng-Granate (high-explosive rocket grenade) 4581 was 1489mm long, carried 125kg of explosives and had a total weight of 345kg. Of the 1400 projectiles ordered, 397 were issued to the weapons office and 317 of those to troop units. The Raketen-Hohlladungs-Granate (hollow-charge rocket grenade) 4592 was introduced for the Kriegsmarine in 1945, which could penetrate 2·5 metres of reinforced concrete.[36]

It is a surprise that the Army Weapons Office, Directorate of Ballistics and Munitions, did not finalise the range table for the RW 61 with Raketen-Spreng-Granate 4581 and Raketen-Hohlladungs-Granate 4592 until 19 November 1944, when it was distributed to the field. It said:

"The provisional range conversion table (Rh. Bo. Ausgabe für 38cm Raketen-Spreng-Granate 4581) - new designation 38cm Raketen-Spreng-Granate 4581 (abbreviated: R Spr Gr 4581) - is supplemented by a corrected table for projectile weight. It should also be noted that the 38cm Raketen-Hohlladungs-Granate 4592 can be fired according to the same table. The Rh.-Bo. Table is to be reprinted as soon as possible after completion and published as a HDv."[37]

On 26 November 1944, the following was noted for the RW 61 with Raketen-Spreng-Granate 4581 and Raketen-Hohlladungs-Granate 4592:

"The new improved impact fuse (with mechanical relays) met all expectations for high sensitivity in demonstration shooting, even in ricocheting up to a 7° angle of descent. The detonator is currently being fired at Wa Prüf (BuM) 1 for acceptance."

On 3 December 1944, the following was noted for both types of projectiles:

"To simplify production, the projectile chamber and cover are being tested as pressed components. Final service instructions and ammunition description are being prepared."

An announcement dated 10 December 1944 stated that the range charts had been provided to the printer as Schußtafel Heeres-Dienstvorschrift 119/988:[38]

"The firing table is intended for both 38cm Raketen-Spreng-Granate 4581 and Raketen-Hohlladungs-Granate 4592. The final firing tables will be at H Vv[39] in the second half of December. Troop units are in possession of a preliminary firing table."

At the same time, improvements to the projectiles were being made, as indicated by the following entry in an announcement dated 17 December 1944:

"Projectiles with non-threaded connection (clamping ring) between the projectile head and propellant and with longitudinally welded shell (formerly seamless tube) are in production. Testing will be carried out after manufacture with powder acceptance firing at Unterlüß."

It was reported that the operating instructions and ammunition description for both projectiles were completed and being issued to the field on 24 December 1944.[40]

The aforementioned powder acceptance firing took place on 19 January 1945 at Unterlüß, but the projectiles delivered did not meet requirements. There was an onset of sputtering when fired at a powder temperature of +/- 62° As a result, there was a considerable reduction in range. A new powder mixture was ordered, and acceptance firing was prepared at Unterlüß and stability tests in Kummersdorf.[41]

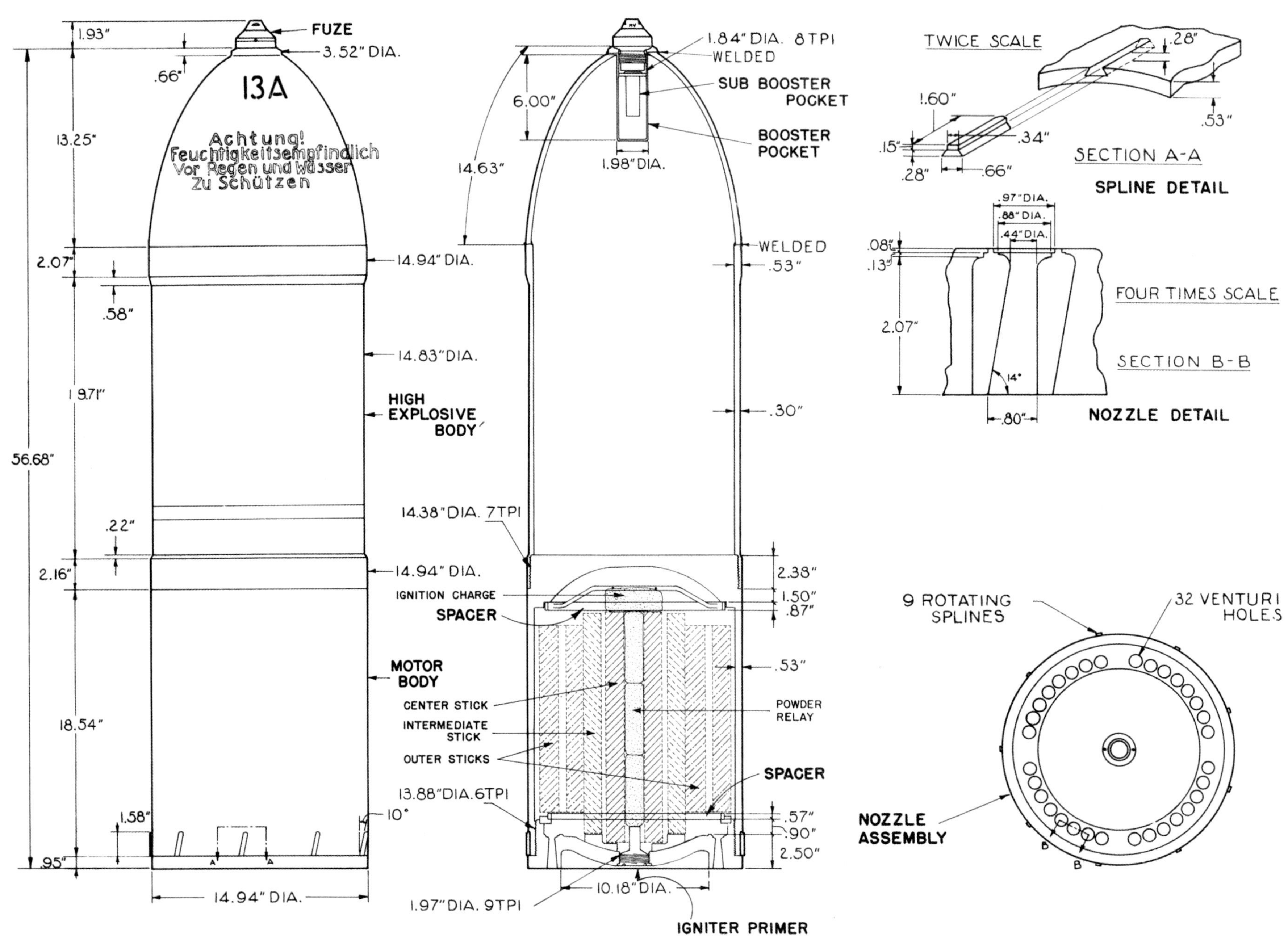
FUZE
1.93"
3.52" DIA.
.66"
13A
Achtung!
Feuchtigkeitsempfindlich
Vor Regen und Wasser
zu Schützen
13.25"
2.07"
.58"
19.71"
56.68"
.22"
2.16"
MOTOR
BODY
18.54"
1.58"
10°
.95"
14.94" DIA.
A   A
1.84" DIA. 8 TPI
WELDED
SUB BOOSTER
POCKET
6.00"
BOOSTER
POCKET
14.63"
1.98" DIA.
14.94" DIA.
WELDED
.53"
14.83" DIA.
.30"
HIGH
EXPLOSIVE
BODY
14.38" DIA. 7 TPI
14.94" DIA.
2.38"
IGNITION CHARGE
1.50"
.87"
SPACER
.53"
CENTER STICK
POWDER
RELAY
INTERMEDIATE
STICK
OUTER STICKS
SPACER
13.88" DIA. 6 TPI
.57"
.90"
2.50"
1.97" DIA. 9 TPI
10.18" DIA.
IGNITER PRIMER
TWICE SCALE
.28"
1.60"
.34"
.53"
.15"
.28"
.66"
SECTION A-A
SPLINE DETAIL
.97" DIA.
.88" DIA.
.44" DIA.
.08"
.13"
FOUR TIMES SCALE
2.07"
SECTION B-B
14°
.80"
NOZZLE DETAIL
9 ROTATING
SPLINES
32 VENTURI
HOLES
B
B
NOZZLE
ASSEMBLY

Raketen-Hohlladungs-Granate 4592

Raketen-Spreng-Granate 4581

Details of the Raketen-Spreng-Granate 4581 from US Technical Intelligence Reports 184 and 192. **Left:** The projectile with finger pointing at the joint between the nose and tail sections. **Top: and bottom:** The contents of the tail section. *3x NARA*

18

From the instruction manual: a perfect side-view. The chassis was given a fresh coat of 'Zimmerit' and camouflage painted with the trackguards fitted. Wet mud has completely covered the running gear. On this side, the tools, yet to be fitted, were a tanker's bar and a spade. Note the four hooks on the side of the engine deck to carry spare tracks. 'Geheim' translates as 'secret'. The following series of photos were probably taken at Sennelager in October 1944. **TTM**

While the sides had fresh 'Zimmerit', the bow armour has missing patches. The tow points with notches, to allow movement of towing and lifting shackles, indicates the chassis was made after January 1944. **TTM**

Possibly Sennelager, Germany
Autumn 1944

How to load ammunition into a Sturm-Mörser. The projectile was transported in wooden shuttering. Here, the crew have rolled it to the right side of the Sturm-Mörser. The stencilling on the shuttering reads: *"Verpackung für 4581 u. 4592"* (Packing for 4581 and 4592). **TTM**

A crewman breaks the steel bands around the shuttering with a tanker's bar. The roadwheels have been camouflage-painted, which would have made them flash when the vehicle was moving. **TTM**

With the shuttering removed, the crew lowered the crane and fastened a strap around the projectile. The white painted band around the centre indicated the optimum point of balance. The small angled splines at the base of the projectile engaged with grooves in the barrel of the weapon. *TTM*

Two of the crew steady the projectile, another lifts it with the crane. At the same time, another gets ready to pivot it into the loading hatch on the roof. Loading ammunition would have been a time-consuming process. *TTM*

This page: The projectile looks like a tight fit for the loading hatch. The main body of the loading hatch was moved using the crane and now rests on the roof. Two of the crew are now inside the vehicle. The two spigots welded to the roof on either side of the hatch are for a rangefinder. **Opposite:** The crew used a roof-mounted winch to load the weapon, the runners of which could be folded away when loading ammunition. Here, a projectile is about to be lowered onto the loading trolley. *2x TTM*

Achtung!
...higkeitsempfindlich
...r Regen und Nässe
zu schützen
TSGEHEIMNIS lt. §88 R STGB
24753 B

Achtung!
...keitsempfindlich
vor Regen und Nässe
zu schützen
STAATSGEHEIMNIS lt. §88 RStGB
24755B

Opposite: Once on the loading trolley, the fuze was screwed into the projectile and had two settings: instant and delayed. **This page:** The breech partially opened. The gunsight mount is to the left; under this is the elevation handwheel. To open the breech, the handle (14) was turned clockwise. The toothed rack (29) engaged with a gear on the handwheel to open the breech. *TTM*

Pages 30-34 show the rest of the firing sequence. **This page:** The projectile was pushed off the trolley (not shown) into the barrel using a jig. The jig engaged with the holes at the back of the projectile, allowing the crew to push it into the breech and rotate it to engage the splines with the barrel's rifling. Immediately below the weapon is the back of the radio operator's seat. **Opposite:** With the projectile in the barrel, a spring-loaded plunger (20) prevented the projectile from falling back. The text *"Nitrocelloidscheibe"* translates as Nitrocelluloid disc. This view shows the two small handwheels on the roof for locking the weapon in the travelling position, and the ventilator in the roof. **Page 32:** The loader slides the firing mechanism (5) up and inserts the igniter. The lack of working space is obvious. **Page 33:** The purpose of the round-handled device is not known, although it may have been used to remove unfired rounds. **Page 34:** With the breech closed, the weapon is ready to be layed in traverse and elevation and then fired. *5x TTM*

20
Nitrocelloidscheibe
STAATSGEHEIMNIS lt. § 88 R STGB
84758 B

5

STAATSGEHEIMNIS lt. §88 RStGB
247608

STAATSGEHEIMNIS H. 888 R STGB
R.11 B
24754B

August 1944

On 1 August, the Polish Resistance Movement began an uprising in Warsaw, which caused additional trouble for the German High Command. The Russian summer offensive had reached the outskirts of Warsaw, whose railyards were important for the Germans as a logistical and reinforcement hub. The German High Command was therefore forced to assign valuable combat units to quell the uprising. At the same time, several ad hoc elements were raised from German occupation forces and the SS. Troops best suited for urban combat: engineers, assault guns and heavy artillery, were also sent to Warsaw to provide heavy weapons support to the infantry. Experience from the Stalingrad disaster had shown that urban combat demanded highly mobile and armoured assault forces to knock out enemy strongpoints. An enclosure to the daily logs of Armee-Ober-Kommando 9, dated 20 August 1944, provides an order of battle of the German forces engaged in Warsaw:

A. Kampfgruppe Rohr
B. Kampfgruppe Reinefarth
    a. Angriffsgruppe Dirlewanger
    b. Angriffsgruppe Reck
    c. Angriffsgruppe Schmidt
    d. Directly attached to Kampfgruppe Reinefarth
        I. Reserves
            1. Sturmgeschütz-Ersatz-Abteilung 200: *(3 Sturmgeschütze)*: 3 officers, 157 enlisted men
            2. Panzer-Abteilung (Fkl) 302: *(20 Sturmgeschütze, 50 radio-controlled tanks)*
            3. Sturm-Pionier-Regiment Herzog mit Pionier-Bataillon 46: 14 officers, 664 enlisted men
            4. Panzer-Pionier-Sturm-Bataillon 500: (minus 1 Kp.): 4 officers, 281 enlisted men
            5. Sturm-Panzer-Abteilung 218: *(10 Sturmpanzers)*: 1 officers, 77 enlisted men
            6. schwere Mörser-Batterie 638: *(1 mortar)*: 3 officers, 110 enlisted men
            7. schwere Stellungs-Mörser-Batterie 201: 2 officers, 62 enlisted men
            8. Sturm-Mörser-Batterie 1000: *(2 38cm mortars on Panzer VI chassis)*: 2 officers, 54 enlisted men
            9. 3./XXI./Polizei-Bataillon Sarnow: 2 officers, 110 enlisted men
            10. Flammen-Werfer-Bataillon Krone: *(150 flamethrowers)*: 10 officers, 292 enlisted men
            11. Feuerschutz-Polizei-Bataillon 96 (mot.): 10 officers, 192 enlisted men
        II. Covering Force
    e. Forces en route[42]

Sturm-Mörser-Batterie 1000, as it was designated on 13 August 1944, was one of the ad hoc forces that did not exist when the uprising started. The table of organization (Kriegsstärkenachweisung or K.St.N.) for a Sturm-Mörser company was not even issued until 15 September 1944. According to K.St.N. 1161, a Sturm-Mörser company comprised:

- 5 officers, 31 NCOs, 43 enlisted men: Total 79
- 4 Panzersturmmörser 38cm 'Tiger'
- 5 heavy machine guns, 34 rifles and 28 pistols
- 1 Sd.Kfz. 251/18, armoured observation halftrack
- 11 trucks
- 4 half-tracked prime movers

The battery was organized into a headquarters section (with the Sd.Kfz. 251/18, an artillery observation variant) and two Sturm-Mörser platoons, with two Sturm-Mörsers each.[43]

The order to establish Sturm-Mörser-Batterie 1000 was issued on 13 August 1944. Initially, the plan was to activate it in the area of the Befehlshaber des Ersatzheeres (Commander of the Replacement Army). Combat readiness was to be established by September, but the situation in Warsaw demanded its immediate deployment. The unit arrived in the Warsaw area in the third week of August, and the unit designation changed from Batterie to Kompanie.[44] All that was available to raise the new battery was the prototype and two Sturm-Mörsers still at Alkett in Berlin. These three vehicles were the only ones approved by the Heeres-Waffenamt (Ordnance Department of the Army) in August 1944 and were directly assigned to the battery. The gravity of the situation can be understood from the following two documents.

A file note from the operations section of Heeresgruppe Mitte, dated 12 August 1944, contains the following:

"Call from Generaloberst Guderian. For the engagement in Warsaw, Army Group will receive as of 14 August:

1. *1 Tiger with 38cm missile launcher (experimental weapon), not suitable for combat against armour-piercing weapons, because structure is made of mild steel*
2. *1 Company with 10 assault guns equipped with 15cm Skoda mortar (2./ Sturm-Panzer-Abteilung z.b.V. 218) [sic]*
3. *1 Radio-controlled tank company*
O.K.H. emphasises engagement of these weapons in street fighting." [45]

Another file note from the operations section of Heeresgruppe Mitte, dated 12 August 1944, states:

"Call from Major Busse, O.K.H. - Operations Section - Subject: Transfer of a Sturm-Mörser battery (experimental weapon).

Early on 13 August, Heeresgruppe Mitte will receive by express train on a special rail car, the first Sturm-Mörser of a battery consisting of three Sturm-Mörsers. Sturm-Mörser: Weight 65 tons on self-propelled chassis; Calibre: 38cm; Ammunition: 6 rounds; Crew: Civilians from the production plant. The remaining 2 Sturm-Mörser will be delivered together with more ammunition over the next few days. A.O.K. 9 received orders by telephone to replace the crews with military personnel." [46]

Generaloberst Guderian was obviously willing to sacrifice the prototype, even though the vehicle was of limited combat value for the mission. It had

to be considered an experimental weapon only. The weapon was so new that the Army was reliant on civilian workers to crew the first vehicles.[47] The prototype arrived in Pruszków on the southwestern outskirts of Warsaw on 12 August 1944, according to an unconfirmed source.[48] Regardless of whether this information is true, the daily transport report of Armee-Ober-Kommando 9 only confirmed the arrival of the other two Sturm-Mörsers on 15 and 17 August at Błonie railyard west of Warsaw.[49] The daily reports mentioned Sturm-Mörsers for the first time in the morning report of 13 August. In this report, it was indicated that one Sturm-Mörser would leave Germany the same day. The next day, the daily report was more precise and provided the information that the Sturm-Mörser, which was supposed to have left Germany the day before, was an experimental tank, i.e. the prototype. In a report dated 14 August, the information was more precise. For the first time, Sturm-Mörser-Batterie 1000 was listed as the unit being sent to Armee-Ober-Kommando 9 with 3 Sturm-Mörsers. The estimated arrival of the first vehicle was 14/15 August. This was the last report mentioning the prototype; all other reports were related to the other two vehicles.[50] The aforementioned unconfirmed source claimed that the prototype moved to Warsaw on 13 August and was engaged against the urban districts of Starowka and Mokotów over the next few days.[51] It is the author's belief that the prototype, regardless of whether it arrived at Pruszków on 13 August, was immediately returned to Germany due to its limited combat value. It should be noted that Albert Speer had sent a message to Adolf Hitler in August 1944, in which he stated that the prototype should be transferred to a new assignment in the West after it had fulfilled its mission in Warsaw. On its way to the West, Speer intended to replace the mild steel superstructure of the prototype at Alkett in Berlin.[52] The daily report of combat-ready tanks and anti-tank weapons in the area of operations of Armee-Ober-Kommando 9 mentioned Sturm-Mörser-Batterie 1000 for the first time on 18 August 1944, when it listed two combat-ready Sturm-Mörsers,[53] which is further proof that no Sturm-Mörsers were engaged in the city until 18 August. The daily logs of the O.K.W. recorded the arrival of the first Sturm-Mörser in Warsaw on 18 August.[54]

Where and when the two Sturm-Mörsers, chassis numbers 250091 and 250230, were between 19 August and the end of the month is not clear. By 24 August, the General Staff of the Army was already requiring the accelerated transmission of an after-action report regarding the engagement of the two vehicles, in order to incorporate the lessons learned into the construction of additional vehicles. This request was sent to the field-army group and the

A Sturm-Mörser in action during the Warsaw Uprising in August 1944. The dark and moody photo emphasises the significant trail left by the projectile, meaning that the vehicle had to move to a different position before firing again. *L.Archer*

Inspector General of Armoured Forces.[55, 56] Heeresgruppe Mitte forwarded this request to Armee-Ober-Kommando 9 the next day.[57] According to the daily logs of the field army, the company was assigned to Kampfgruppe Reinefarth until the end of the month. Except for 20 and 21 August, when it did not submit a report, Sturm-Mörser-Kompanie 1000 consistently reported 2 combat-ready vehicles.[58]

In the meantime, the prototype had been sent back to Germany, and a new crisis developed in the West. The Allies had broken out from Normandy and Paris was about to be liberated. This time, there was no alternative to the prototype, and without sufficient time to up-armour the vehicle at Alkett, it was sent to Paris on its own. By the end of August 1944, the situation in France had developed much faster than the Germans had expected. This forced the unloading of the prototype at Meaux on 25 August 1944, some 50 kilometres outside of Paris. Instead of its use in Paris, the sole combat vehicle of Sturm-Mörser-Kompanie 1000 in the West was engaged at Vouzier on 31 August, according to the combat history of the Panzer-Lehr-Division.[59, 60, 61]

September 1944

In an overview dated 1 September 1944, the General der Panzertruppen (General of the Armoured Forces) reported that Sturm-Mörser-Kompanie 1000 was to be activated through the Replacement Army. This was to occur in the area of the Befehlshaber des Ersatzheeres (Supreme Commander of the Replacement Army), and was to be combat-ready by September 1944. The report did not mention the deployment of the two Sturm-Mörsers in Warsaw or the prototype in the West.[62]

On 2 September 1944, Armee-Ober-Kommando 9 sent the after-action report it had been tasked to produce to Heeresgruppe Mitte. Entitled *"Sturmmörser,"* which stated:

"Sturm-Mörser-Kompanie 1000, which had been engaged in street and house fighting with 2 Sturmmörser on Panzer VI chassis in Warsaw since 19 August 1944, has made the following observations:

In general: So far, the company has only been engaged in street and house fighting. The crews should have been trained on the new type of

gun first.

Tactical opportunities for engagement: The Sturmmörser stood out during the fighting in Warsaw

*a) in overpowering pockets of resistance behind walls and concrete
b) in firing on enemy concentrations and assembly areas due to the heavy and far-reaching impact of shrapnel (500-800 metres)
c) in blowing a breach in barricades of all kinds, in walls and in concrete in close-support of assault detachments*

Due to the slow rate of fire (4 rounds per hour), the company was used to prepare targets for attack and assaults.

Firing positions and moving into position: If possible, the firing position must be reconnoitred beforehand. The target must be fixed, and the distance to the target must be clear. Preparations for firing must be made under cover in order to shorten the time the vehicle is exposed to enemy fire. It is necessary for the vehicle commander to orient himself before moving into the firing position. In order to shorten the time in the firing position even more, it has proven effective to load the weapon while under cover and to move into the firing position when ready to fire.

Target accuracy: The rocket-like projectile increases its initial speed during the first part of its flight path. Range variation is considerable when firing short distances. For this reason and to avoid exposing the vehicle to enemy fire at short range, it is inappropriate to move too close to the target. Excellent target accuracy was achieved at a range of around 600 metres.

Impact: Worthwhile targets were houses with 2 to 3 floors, which were destroyed with a single round. Holes the size of a barn door were made in concrete buildings. The effect on material and morale could be raised by increasing the speed of the firing sequence and by engaging two or more weapons on one target.

Basic ammunition load: With a rate of fire of only four rounds an hour, caused by the difficult loading process, and a capacity of only 14 rounds, it should be checked whether using the valuable chassis of a 'Tiger' is justified. The chassis of a Panzer IV, combined with an armoured ammunition carrier, would fulfil the same purpose, particularly when considering that using the vehicle in the front lines is out of the question due to the slow rate of fire.

Technical issues: Due to technical reasons - maintenance and spare parts supply - an allocation [of these vehicles] to other 'Tiger' units should be considered." [63]

Throughout September, Sturm-Mörser-Kompanie 1000 reported two combat-ready vehicles in Warsaw. The only change took place when Kampfgruppe Reinefarth was attached to Gruppe von dem Bach on 13 September 1944.[64] According to this order, Sturm-Mörser-Kompanie 1000 remained part of Reinefarth's forces, which also had the following artillery units attached:

- Schwere Stellungs-Nebel-Werfer-Batterie 201
- Heeres-Artillerie-Batterie 428
- Heeres-Artillerie-Batterie 638

Kampfgruppe Reinefarth received orders to eliminate the last pockets of resistance in the centre of the city and to clear the western bank of the River Vistula.[65]

On 26 September, planning began for Operation 'Wolfsangel,' which was designed to eliminate the pocket at Żoliborz[66] by the XXXXVI. Panzer-Korps. Sturm-Mörser-Kompanie 1000 was mentioned again in an order by 19. Panzer-Division, dated 28 September (Divisional Order No. 41 for Staging and Attacking Żoliborz). On 27 September, the leader of the insurgents had been asked to surrender, which he refused.

As a result, 19.Panzer-Division was tasked with attacking Żoliborz with three assault groups in a concentric attack on 29 September. To that end, it was ordered to reinforce the division with the following units from Korpsgruppe von dem Bach: Regiment Schmidt, Gruppe Dirlewanger, Panzer-Abteilung (Fkl.) 302, Sturm-Panzer-Kompanie 218 and Sturm-Mörser-Kompanie 1000.[67] Gruppe Schmidt was to attack the southern district. The following elements were attached to Oberst Schmidt and his assault group:

- Regiment Schmidt
- Regiment Reck
- Pionier-Sturm-Bataillon 501
- Kosaken-Abteilung 69
- Panzer-Abteilung (Fkl.) 302
- Sturm-Panzer-Kompanie 218
- Sturm-Mörser-Kompanie 1000 [68, 69]

The attack on the pocket was preceded by an artillery barrage from 06:57 to 07:00. The barrage was under the control of Panzer-Artillerie-Regiment 19 of 19.Panzer-Division, which had the following organic and attached weaponry available: 1x 60cm 'Karl' siege mortar; 12x 21cm guns; 31x heavy field howitzers; 4x heavy cannons; 21x light field howitzers; 15x 8·8cm artillery/anti-tank guns; all available infantry guns; 20x 28/32cm rocket launchers; 38cm Sturm-Mörser; and Heeres-Flak-Abteilung 272. During the second barrage, on 30 September, the Sturm-Mörser and the rocket launchers no longer participated, probably due to their general inaccuracy. During the course of the day, the artillery progressively stopped firing into the pocket so as not to endanger the attacking forces.[70]

Until at least 6 September 1944, the second platoon was still engaged in the West, according to a report of the General of the Armoured Forces.[71]

In August 1944, Albert Speer had reported to Adolf Hitler that Alkett intended to finish seven more Sturm-Mörsers between 15 and 21 September. In the middle of September, he sent Hitler another report which stated that a total of ten Sturm-Mörsers had been assembled by Alkett in September under the guidance of Obermeister Hahne. This brought the overall number of completed vehicles to 13, including the prototype. Hitler ordered its continued production at five a month, but for unknown reasons, this order was only seen through once, in December 1944.[72] The new Sturm-Mörsers were accepted by the Heeres-Waffenamt in September, but were not assigned to a formation.[73] Nevertheless, the availability of additional Sturm-Mörsers required the activation of further companies. On 3 September 1944, Sturm-Mörser-Kompanie 1001 was ordered to establish combat readiness by 23 September.[74] On 27 September, another report from the General of the Armoured Forces showed another Sturm-Mörser company, numbered 1002, in the process of activation.[75]

October 1944

In October, the 10 Sturm-Mörsers manufactured in September were finally assigned to the General of Armoured Forces, who sent them to Sennelager near Paderborn, and became the central issuing point for Sturm-Mörsers.[76] Six vehicles were assigned to Sturm-Mörser-Kompanie 1000 and 1001 in October. These were chassis numbers 250237 and 251168 (both newly assigned to Sturm-Mörser-Kompanie 1000) and 250043, 250051, 250103 and 250471 (all assigned to Sturm-Mörser-Kompanie 1001). One vehicle, chassis number 251174, was assigned to Panzer-Ersatz- und Ausbildungs-Abteilung 500. Another, chassis number 250237, was sent to the Rheinmetall-Borsig factory for examination and approval by the Heeres-Waffenamt, indicating it had developed technical problems. Two vehicles, chassis numbers unknown, were retained by the General of Armoured Forces. It is likely that both were intended for Sturm-Mörser-Kompanie 1002, which was in an early state of activation at that time.[77, 78] Starting October 1944, Panzer-Ersatz- und Ausbildungs-Abteilung 500 had become the replacement unit for all Sturm-Mörser units, so it made sense to assign two vehicles to them for training purposes.[79, 80]

The one platoon of Sturm-Mörser-Kompanie 1000 in Warsaw remained attached to Gruppe von dem Bach until 7 October. It was then attached to the XXXXVI. Panzer-Korps until 15 October, when orders arrived to move the company to Budapest. Until this point, they had repeatedly reported two combat-ready Sturm-Mörsers. The company finally left Warsaw at 19:40 on 17 October. On 19 October, orders arrived to return to Warsaw from Budapest. The next morning, at 00:10, they unloaded at Warsaw with only one Sturm-Mörser. The company was attached to the XXXXVI. Panzer-Korps again. Until the end of the month, Armee-Ober-Kommando 9 continued to report only one combat-ready Sturm-Mörser.[81, 82, 83, 84]

The company's transfer to Budapest was obviously related to the German-instigated uprising against the Horthy government which started on 15 October, to prevent the Hungarians from leaving the alliance and joining the Russians. Surprisingly, it was over by 17 October, as the Horthy government offered no resistance. There was no longer a need for the Sturm-Mörsers, which had just left Warsaw. When and where the information to return to Warsaw reached the company remains unclear. What really happened during their 52-hour absence is also unclear and will remain a mystery due to a lack of reliable sources.

There are rumours that it had been intended to attach the company to Panzer-Brigade 109 in the area of Budapest, but Panzer-Brigade 109 was engaged in the Debrecen area by the time the company was on its way to Budapest, i.e. about 240 kilometres east of the Hungarian capital. By then, orders were in place to consolidate Panzer-Brigade 109 with Panzer-Grenadier-Division 'Feldherrnhalle'. None of these orders showed a link between the brigade and Sturm-Mörser-Kompanie 1000.

To add to the confusion, there was a statement that the Russians had captured a Sturm-Mörser in Hungary in 1944, in the village of Nyékládháza, almost 100 kilometres northwest of Debrecen and 180 kilometres east of Budapest. Nyékládháza is located on the railway line from Budapest to Warsaw (via Miskolc), which at least might explain how the vehicle could have come to the village. By the time the transport passed through Nyékládháza on 18 or 19 October, the village was far behind the front, making the loss of this Sturm-Mörser doubtful.

There can be no doubt that the Sturm-Mörser in question would have been one of the two vehicles from Sturm-Mörser-Kompanie 1000, with the chassis number 250091 or 250230. Images of the captured vehicle verify that it is the one on display at the Patriot Park in Kubinka, Russia. Recent research made it clear, however, that the images of the Sturm-Mörser were not taken at Nyékládháza, but in Germany in May 1945.

As a result, we can only guess why only one Sturm-Mörser returned to Warsaw. We know that one was assigned to the Inspector General of the Armoured Forces in December, which brought the overall number of assigned vehicles to eleven. Remember that the prototype and first two vehicles had not been assigned to the Inspector General. The Inspector General's December report also showed that no Sturm-Mörsers had been lost up to December 1944. The author believes that the missing vehicle developed serious mechanical problems and was sent to Germany by Sturm-Mörser-Kompanie 1000 for an overhaul. Once the vehicle had been repaired; it was then assigned to the Inspector General of the Armoured Forces.

On 22 October, an order was issued to withdraw the Company for rest and refitting in the operational area of the Commander-in-Chief of the Replacement Army.[85] Sturm-Mörser-Kompanie 1000 was still in Warsaw with one platoon, when a second platoon was activated at Sennelager after the platoon equipped with the prototype had returned from France. Two Sturm-Mörsers were assigned to the company on 26 October,[86] while the prototype was reissued to Panzer-Ersatz- und Ausbildungs-Abteilung 500.[87]

In correspondence from Waffen-Prüf-Amt to Albert Speer on 10 January 1945, the operational experiences of Sturm-Mörser-Kompanie 1000 in Warsaw were once again addressed:

"The Panzer-Sturm-Mörser on a Tiger chassis has hitherto been employed with only two vehicles in the Warsaw street fighting, where it was particularly suited due to its heavy armour. Operations demonstrated the advantages of the vehicle in combatting large individual targets, such as barricades and buildings. The long loading time was disadvantageous as the crew was overexposed to enemy fire." [88]

Status reports for Sturm-Mörser-Kompanie 1001 from early October state that the company was raised from the Replacement Army. It was activated at Sennelager, the home of Panzer-Ersatz- und Ausbildungs-Abteilung 500, which was the replacement unit for all units equipped with 'Tiger' tanks. The completion of activation was anticipated for the end of October. Despite this, the company did not get its Sd.Kfz. 251/18 artillery observation halftrack, until sometime in November (issued from the depot on 3 November). On 4 November, four Sturm-Mörsers were allocated to the unit and dispatched the same day. For quite some time, there were plans to send the company to Warsaw as well. The files of the transport officer of Armee-Ober-Kommando 9 show that as of 25 October 1944, it was planned to transfer the company from Sennelager to Warsaw in early November. This information was repeated in the daily reports until 11 November 1944.[89, 90, 91]

On 7 October, a document from the directorate of the General der Panzertruppen mentioned that Sturm-Mörser-Kompanie 1002 was also to be activated from the Replacement Army in the area of operations of the Commander-in-Chief of the Replacement Army.[92]

By the end of October 1944, the 13 Sturm-Mörsers were assigned as follows:

Stu.Mrs.Kp.1000	4
Stu.Mrs.Kp.1001	4
Pz.Ers. u. Ausb.Abt.500	2
HZa Sennelager	2
Borsig/Waffenamt	1

November 1944

In late Autumn 1944, the planning for Operation 'Herbstnebel', the German Ardennes Counteroffensive, led to a constant build-up of forces in the Eifel region and west of Cologne. The build-up in the Eifel region was carried out under total secrecy to disguise the German plans. The build-up of forces west of Cologne and north of the Eifel was executed under normal conditions to make the Allies believe the Germans were gathering a force to meet the Allied push towards the Rhine east of Aachen. In a report by the O.K.W. dated 23 November 1944, Sturm-Mörser-Kompanie 1000 and 1001 were assigned to OB West (Oberbefehlshaber West = Commander-in-Chief West) for the upcoming counteroffensive.[93] By this time, Sturm-Mörser-Kompanie 1001 had already arrived in the West.

As early as 2 November 1944, Heeresgruppe B (Army Group B) had requested the movement of Sturm-Mörser-Kompanie 1001 to Zülpich. The Field Army Group intended to assemble the Sturm-Mörsers in the shortened line near Heimbach.[94] Three days later, a report by the General der Panzertruppen stated that the company was in the West with four combat-ready vehicles.[95] Another report by the Oberbefehlshaber West (Commander-in-Chief West) dated 7 November stated that the company had been en route to Zülpich since 5 November.[96] On 10 November, an overview showing General Headquarters, Field Army, and Corps armoured units claimed that the company was in the West with 4 Sturm-Mörsers.[97] All these reports were misleading, because it was not until 13 November that the Transport Officer of Armee-Ober-Kommando 7 reported the arrival of Sturm-Mörser-Kompanie 1001 at Zülpich at 15:25.[98] Two days later, OB West and Heeresgruppe B both reported that the company had arrived at Zülpich by train.[99, 100]

The Americans also noticed the arrival of the company, although it remains doubtful they were fully aware of this secret new German weapon. In the last week of November, they captured Leutnant Schutkowski, an officer of the 47.Volks-Grenadier-Division. When interrogated, he stated he had heard from a railway official at Düren, that a train was unloaded on 15 November, with six specially modified Tiger I tanks mounting 48cm mortars. Although the number of assault tanks and the calibre of the gun were obviously wrong, the official had observed the arrival of Sturm-Mörser-Kompanie 1001 in the Heeresgruppe B area.[101] It remains unclear where the company went after detraining at Zülpich. A secondary source mentions that it assembled near Gemünd.[102] After the war, the Commanding General of the LXXXI. Armee-Korps, General Köchling, claimed that one Sturm-Mörser company was attached to his corps in mid-November. He went on to say that later on - until early April 1945 - two Sturm-Mörser companies were temporarily attached to his corps.[103, 104]

Sturm-Mörser-Kompanie 1000 was expected to arrive in the Heeresgruppe B area between 20 and 25 November 1944. It arrived on 23 November, according to an OB West report.[105, 106, 107] Reports of the General der Panzertruppen, dated 5 and 10 November, stated the company was in the Homeland War Zone for rest and refitting with four combat-ready Sturm-Mörsers, but this information is again misleading.[108, 109] As of 20 November 1944, the daily strength reports of Armee-Ober-Kommando 9 continued to report that the company was still attached to the XXXXVI. Panzer-Korps in Warsaw with one combat-ready Sturm-Mörser.[110, 111] This single vehicle remained in Warsaw until 22 November, when it was finally loaded onto train number 231295, heading for the Sennelager training area.[112] The departure of Sturm-Mörser-Kompanie 1000 on 22 November was also noted in a report by the XXXXVI. Panzer-Korps from 24 November concerning attached units. This information appeared under the heading of 'departed.'[113] According to an overview dated 1 December 1944, one of the Sturm-Mörsers that had remained in the inventory of the Generalinspekteur der Panzertruppen at Sennelager was assigned to one of the three Sturm-Mörser companies in November. It is the author's belief that this vehicle was allocated to Sturm-Mörser-Kompanie 1000, which was short one combat vehicle after the deployment to Hungary in October 1944.[114]

There were some conflicting reports on Sturm-Mörser-Kompanie 1002 in November. Two reports dated 2 November claimed that the company was to receive its full complement of tracked and wheeled vehicles by 5 November. It was also to reach full combat readiness the same day. During this period, the company was still at Sennelager. On 5 November, the company was reported in the homeland war zone, but still minus

Sturm-Mörsers. On 10 November, a report added the information that Panzer-Ersatz- und Ausbildungs-Abteilung 500 be charged with providing personnel for the activation of the company. By the end of the month, Sturm-Mörser-Kompanie 1002 was still at Sennelager without tracked and wheeled vehicles.[115, 116, 117, 118]

By the end of November 1944, the 13 Sturm-Mörsers were allocated as follows:

Stu.Mrs.Kp.1000	4
Stu.Mrs.Kp.1001	4
Pz.Ers. u. Ausb.Abt.500	2
HZa Sennelager	2
Borsig/Waffenamt	1

December 1944

Reports regarding Sturm-Mörser-Kompanie 1000 were again conflicting at the start of the month. According to the General der Panzertruppen, two Sturm-Mörsers, which had been issued on 26 October 1944, were finally transported to Sennelager on 1 December.[119] This seems unusual inasmuch as the two Sturm-Mörsers had been assigned to Army Depot Senne since October. This is also in contrast to the November reports, which had repeatedly stated that the company had been issued the full complement of four Sturm-Mörsers. However, on 3 December, the General der Panzertruppen reported that the company was still at Sennelager for rest and refitting. All wheeled vehicles were said to be available, but the combat vehicles were not due until 10 December. Nevertheless, it was expected that the company would be combat ready by 7 December.[120] As previously mentioned, the company was attached to Heeresgruppe B to take part in the upcoming Ardennes offensive. By 10 December, its arrival was already overdue in the west. The same day, the Oberbefehlshaber West sent a message to Heeresgruppe B at 13:30, which stated that the arrival of the company is no longer necessary at all costs but, at the same time, it confirmed that the company was in transit to the west.[121] On the same day, the General of the Armoured Forces reported the company's arrival in the west with four combat-ready Sturm-Mörsers.[122] It would take another five days before Sturm-Mörser-Kompanie 1000 arrived at Elsdorf from Sennelager (15 December). A message from Heeresgruppe B[123] stated that the company arrived in its area of operations with 2 officers, 17 noncommissioned officers, 66 enlisted men and 4 Sturm-Mörsers.[124, 125] Except for a note from the General der Panzertruppen, which reported Sturm-Mörser-Kompanie 1001 in the west with four combat-ready vehicles on 10 December,[126] there are no other reports that mention the company before 15 December 1944. The day before the start of the Ardennes offensive, both Sturm-Mörser companies were officially attached to Armee-Ober-Kommando 15.[127, 128] Later that day, the Oberbefehlshaber West forwarded an order issued from Hitler which read:

"For the expected heavy defence in the Liege area, the left flank of the LXVII. Armee-Korps will be reinforced by the following heavy batteries:

Sturm-Mörser-Kompanie 1000 and 1001
Schwere Batterie 428 and 628 (Karl)
An unidentified 21cm Granatwerfer-Kompanie" [129]

Both Sturm-Mörser companies were assigned to the LXVII. Armee-Korps on 16 December after the offensive had started. The corps was engaged on the right flank of the 6.Panzer-Armee with the mission of guarding it against enemy reinforcements from the north on the line of Kesternich, Eupen and Liège.[130, 131] It appears that the companies did not move forward on 16 December. Reports mention both in the Gemünd-Elsdorf area with seven combat-ready Sturm-Mörsers. One vehicle from Sturm-Mörser-Kompanie 1000 had obviously broken down.[132, 133] So far, no reports have come to light that confirm the engagement of the companies in the LXVII. Armee-Korps area in December 1944. One report confirms that one company had moved closer to the front. At 01:25 on 18 December 1944, the officer on duty at AOK 15 reported that Sturm-Mörser-Kompanie 1000 had been located in the area of the LXVII. Armee-Korps. The ammunition transports were ordered to march in the direction of Gemünd. The leader of this transport was instructed to move ahead to the corps headquarters at Dalbenden.[134] According to the situation maps of the Oberbefehlshaber West ('Lage Frankreich' = 'Situation: France'), Sturm-Mörser-Kompanie 1000 reported 3 combat-ready Sturm-Mörsers through 26 December; thereafter and until the end of the year, the number was four. Sturm-Mörser-Kompanie 1001 reported 4 combat-ready Sturm-Mörsers between 22 and 31 December 1944.[135]

By the end of the month, Sturm-Mörser-Kompanie 1000 and 1001 were in the west with four combat-ready Sturm-Mörsers each, while Sturm-Mörser-Kompanie 1002 was still at Sennelager, waiting to finish activation.[136, 137, 138]

Sturm-Mörser-Kompanie 1002 had spent another month at Sennelager. On 3 December 1944, it was still intended that all tracked and wheeled vehicles be received by 15 December and to establish combat readiness by 20 December.[139] On 8 December, two Sturm-Mörsers were finally issued. These were chassis number 250237, previously sent to Borsig for further testing and evaluation, and chassis number 251174, previously issued to Panzer-Ersatz- und Ausbildungs-Abteilung 500.[140, 141] On 13 December, two more Sturm-Mörsers were issued. It is likely that one was the remaining vehicle at Army Depot Senne.[142] While the prototype was still issued to Panzer-Ersatz- und Ausbildungs-Abteilung 500 at Sennelager,[143] the second must have been assigned from the pool of newly manufactured vehicles. It is worth mentioning that a final batch of five Sturm-Mörsers was completed by Alkett and approved by the Heeres-Waffenamt in December 1944.[144] Nevertheless, it is unclear as to why the General der Panzertruppen continued to report on 10 and 25 December, that Sturm-Mörser-Kompanie 1002 was at Sennelager without Sturm-Mörsers.[145, 146]

By the end of December 1944, the 18 Sturm-Mörsers were assigned as follows:

Stu.Mrs.Kp.1000	4
Stu.Mrs.Kp.1001	4
Stu.Mrs.Kp.1002	4
Pz.Ers. u. Ausb.Abt.500	1
Waffenamt	5

January 1945

No sources have been found that prove the assumption that Sturm-Mörser-Kompanie 1000 and 1001 saw action in the Ardennes. Two secondary sources claim that Sturm-Mörser-Kompanie 1000 was in the Trier and Alsace area at the end of December 1944 / early January 1945, but this information is clearly wrong.[147, 148] Until 19 January 1945, both companies each reported four combat-ready vehicles on a daily basis.[149] American intelligence reports did not mention any super-heavy artillery fire that could be traced back to the Sturm-Mörsers. By the middle of January, the Eifel front was almost back to where it was when the German offensive started on 16 December 1944. After a month of bitter fighting and significant losses in manpower and materiel, the German front line was thinned out. The old Westwall line, although of limited military value in 1945, at least provided shelter for the exhausted German troops. For over a month, the front came to a standstill.

The Americans prepared for the final attack into the Reich, reinforced their troops, stockpiled supplies, and started to reconnoitre German positions along the Eifel - Roer riverfront. It was in this period that the two Sturm-Mörser companies saw their first engagements. A prisoner from Sturm-Mörser-Kompanie 1001, captured by the Americans in May 1945, stated that his company was in action along the Roer river from January until March 1945. During this period, the company was said to have fired 47 rounds on American positions.[150]

Heinz Matten confirmed this statement after the war. Matten, a former member of Sturm-Mörser-Kompanie 1001, said that the company had three combat-ready vehicles in January and February 1945 in the Düren - Urft Valley area. Starting 20 January, Sturm-Mörser-Kompanie 1001 reported only three combat ready vehicles, which makes it simple to confirm Matten's statement.

He continued that the company had considerable success, and allegedly destroyed a large Westwall bunker with a single rocket that was occupied by the enemy. In another action, they fired a rocket into a small village occupied by the Americans. Almost all the 'Sherman' tanks in the village were put out of action, and the crews killed or wounded. Excellent effects were also witnessed when rockets were fired on dug-in infantry and artillery positions. Matten also claimed that the Americans were paralysed for hours after they had been shelled by the company. Later on, the Americans reportedly started to hunt down the Sturm-Mörsers with low flying aircraft, spies and deserters. Often, several enemy artillery battalions fired on a Sturm-Mörser when spotted. This forced the Sturm-Mörsers to remain in a firing position for only a few minutes and to change position after firing. Later on, one Sturm-Mörser broke down and had only limited use.[151]

Leutnant Doll was an officer in Sturm-Mörser-Kompanie 1001. Without mentioning a specific date, he recalled an engagement west of the Rhine, where a single rocket destroyed an enemy tank assembly area in a village. He referred to the impact as disastrous.[152] At the end of January, a German prisoner reported[153] to have seen a large assault mortar near Hambach, east of Jülich. The vehicle probably carried 12 projectiles, of which 10 were fired on the American positions.[154]

Very few official reports mention the three Sturm-Mörser companies during January. There were the daily strength reports of the Oberbefehlshaber West that ended on 23 January. There were also the daily maps of the Supreme Commander West, which clearly show that Sturm-Mörser-Kompanies 1000 and 1001 were attached to the LXXIV. Armee-Korps from 20 January until at least 27 February. Another report, dated 6 January, mentioned that Sturm-Mörser-Kompanie 1002 still was at Sennelager finishing activation after most of its combat and other vehicles had been issued in December. A General der Panzertruppen report showed that Sturm-Mörser-Kompanie 1000 and 1002 finally received their missing Sd.Kfz. 251/18 on 18 January.[155, 156, 157, 158]

On 19 January, the Abteilung Ausbildung at the O.K.H. concluded that separate 'Tiger' Sturmmörser units were neither technically nor tactically viable. An immediate consolidation into 'Tiger' battalions as the fourth company was required.[159] The recommendation came two days too late, as Hitler had ordered the immediate transfer of the Sturm-Mörser companies from the armour to the artillery branch on 17 January. The General Staff of the Army was asked to issue the necessary implementation regulations in coordination with the Inspector General of Armoured Forces. The three Sturm-Mörser companies were ordered to raise the number of assigned combat vehicles from four to six. Implementation was the responsibility of the General Staff of the Army.[160]

On 23 January 1945, the O.K.H. arranged to hand over all files and relevant information from the Inspector General of Armoured Forces to the General of Artillery at the O.K.H. Effective immediately, the three Sturm-Mörser companies were redesignated as 'Sturm-Mörser-Batterien (RW)'. The branch of service colour was changed from pink to red. The Commander-in-Chief of the Replacement Army was asked to issue the necessary implementation regulations for his area of responsibility. This was necessary because of the requirement to transfer specially trained personnel and training material from Panzer-Ersatz- und Ausbildungs-Abteilung 500 at Sennelager to Panzer-Haubitzen-Ersatz- und Ausbildungs-Abteilung 201, the new replacement unit for the Sturm-Mörser batteries.[161]

It is not clear if this change was actually carried out. Soldiers of all three Sturm-Mörser units that were captured between February and May 1945 said they belonged to either a company or a battery. Sturm-Mörser-Batterie 1002 returned to Sennelager for rest and refitting in March after its first engagement in the west. New Sturm-Mörsers were also still assigned from Army Depot Senne until the Americans captured it. To simplify matters, the author will continue to refer to the units as companies.

By the end of January 1945, the 18 Sturm-Mörsers were assigned as follows:

Stu.Mrs.Kp.1000	4
Stu.Mrs.Kp.1001	4
Stu.Mrs.Kp.1002	4
Pz.Ers. u. Ausb.Abt.500	1
HZa Sennelager	5

February 1945

In early February, Sturm-Mörser-Kompanies 1000 and 1001 were still in the Eifel - Roer River region. Both companies reported having four Sturm-Mörsers on 5 February, but while Sturm-Mörser-Kompanie 1000 reported all of them as combat-ready, Sturm-Mörser-Kompanie 1001 reported only three as combat-ready.[162] On 23 February, the Commander-in-Chief of the Replacement Army requested the Commander-in-Chief West raise the number of Sturm-Mörsers for Sturm-Mörser-Kompanie 1000[163] from four to six. Army Depot Senne was ordered to provide the two vehicles. Sturm-Artillerie-Schule Burg received the order to send an acceptance detail for the two to Sennelager. The personnel strength of this detail was established at seven noncommissioned officers and four enlisted men. The final issue of the vehicles and transfer of personnel to Sturm-Mörser-Kompanie 1000 was to take place by separate order.[164]

In a supplemental order of 12 February, to the original of 23 January, the Commander-in-Chief of the Replacement Army directed two senior noncommissioned officers from Panzer-Ersatz- und Ausbildungs-Abteilung 500 be transferred to the Artillery Inspectorate. Both were to be familiar with the Sturm-Mörser (38cm R.W. 61). In addition, all of the replacement crews at the replacement and training battalion, along with all additional personnel that had been trained on the Sturm-Mörser, were to be transferred to the artillery. Furthermore, all Sturm-Mörsers being manufactured, and all future vehicles and munitions were to be transferred. The new replacement unit for the three companies was to be the Sturm-Artillerie-Schule Burg (Magdeburg). The Sturm-Mörser-Kompanien were to be immediately redesignated as Sturm-Mörser-Batterien (R.W. 61).

The Oberembt Sturm-Mörser

When the Americans launched their Roer River offensive on 23 February 1945, Sturm-Mörser-Kompanie 1000 was in the area of Vettweiß, southeast of Düren. Thanks to a German deserter from a Volkssturm unit and a local from Vettweiß, who was captured by the Americans around 28 February, some interesting details regarding Sturm-Mörser-Kompanie 1000 have been preserved. On 25 February, the deserter saw two huge self-propelled mortars. He had no idea as to the calibre of the weapons, which were mounted on a specially reinforced King Tiger tank chassis. He had been told by one of the crews that the shell weighed 700 pounds. The shells were loaded into the mortar carriers by a system of blocks and tackles, pulleys and cranes. The crewman also told him that the mortars had been used in the siege of Warsaw,[165] and that they were designed as support weapons with a maximum range of 12,000 yards.

The two vehicles in Vettweiß were what remained of a battery of four, the only such battery on this front. One of the weapons had been damaged, and another was out of commission, needing a new barrel. The last two remained in Vettweiß, except for sorties to the bank of the River Roer to fire off a few rounds. They should have withdrawn earlier, but had run out of fuel. Evidently, enough fuel arrived to take them to Horrem, as that was their destination, according to the deserter's chatty friend on the Sturm-Mörser.[166]

The next day, one Sturm-Mörser took cover against enemy fighter-bombers at the villa near Wissersheim. The commander, a small agile Leutnant, waited until dusk before continuing his move.[167] While one of the two Sturm-Mörsers only made it halfway to Horrem on 26 February, the other must have reached Horrem in the morning, at which point it was immediately redirected to Oberembt, northwest of Elsdorf, and ordered to attack advancing American forces.

The US 30th Infantry Division planned to attack Elsdorf from the south on 26 February. The forward elements had just attacked Giesendorf, south of Elsdorf, when they were hit by the largest calibre ammunition that they had experienced. The first round completely demolished a house, leaving a deep brick-filled hole. Shrapnel flew a hundred yards, knocking everyone within that distance off their feet. Shrapnel found targets; one killing Lieutenant Harrington and another mortally wounding Captain Peters of the 391st Field Artillery. An infantry company commander was killed, Lieutenants Paulsen and Jones were wounded, as were a number of enlisted men. The assumption was that it was a 380 millimetre railway gun.[168] Later that day, it became clear that the shelling had come from the only Sturm-Mörser of Sturm-Mörser-Kompanie 1000, which had moved into a firing position in Oberembt that morning. The vehicle and crew soon met their fate, when the tank tried to move into a new firing position and threw a track. When Company C of the 117th Infantry (30th Infantry Division) attacked Oberembt, it immobilised the Sturm-Mörser that was ready to fire its next round but was facing the wrong way. Although an easy target, after throwing the right track, the Americans attacked the tank from the rear and accompanying tank destroyers fired several rounds into its rear, and the crew either killed or captured.[169] The surviving crew stated that their Sturm-Mörser was one of four vehicles assigned to Sturm-Mörser-Kompanie 1000.[170, 171]

In a newspaper interview in April 2006, Christian Müller from Oberembt, a 13-year-old in 1944, remembered that the crew of four desperately tried to recover their vehicle after the driver had driven the vehicle into a ditch in front of his parent's home, causing it to throw the right track. The crew even considered blowing the Sturm-Mörser up, which would have caused considerable damage to the small village due to the high explosive projectiles inside the vehicle. It had broken down on Neusser Straße, at the north-eastern edge of Oberembt.

One of the crew removed a machine gun from the vehicle and set it up it in the direction of Finkelbach. His defence was in vain: the advancing American infantry simply destroyed the position with a hand grenade, killing the German. The rest of the crew immediately took cover in a nearby cellar, with Christian Müller and two families. When the firing subsided, the Americans demanded that the Germans surrender. Günter Stritzke, one of the crew in the cellar, tried to escape. Although the Americans shouted at him to stop, he did not, and was shot near his vehicle. The Americans were so annoyed that they prevented the burial of Stritzke and 7 or 8 other dead bodies for more than a week.

Müller also recalled that for nearly a month countless Americans came to examine and photograph the vehicle. He climbed onto it once and counted eight projectiles still inside. After a month, the Americans brought several engineer tanks onto Jülicher Straße, and dragged the Sturm-Mörser out of the ditch to the main road, where it was loaded onto a flatbed trailer.[172] It was eventually shipped to the UK for further examination. The Sturm-Mörser was scrapped after the war, except for its gun barrel, which is now on display at The Tank Museum in Bovington, UK.

This first Sturm-Mörser interested the Allies. One of the first reports was issued by the 30th Infantry Division on 26 February:

> *"The chassis reveals some modifications similar to those made in the 'Jagdpanther' tank. It has about 5 inches of armour plate at about a 50-degree angle on the front. The howitzer is situated slightly left of the centre of the tank as the observer faces it. The barrel is 7 feet long and is rifled. The base of the shell is 15 inches in diameter, 5 feet long, and weighs 770 pounds. The ammunition load is placed inside the tank by use of a crane. Inside the tank there is a hoist used to place the shell in position for loading. There is a slide inside the tank to feed the shell into the chamber. There is no recoil mechanism on the weapon, but a heavy sleeve. The howitzer is believed to be fired by a propelling charge inside the shell, and it appears to have some aspects of a rocket. It is believed to have a comparatively short range of 5000-6000 meters. The vehicle is driven with the aid of periscopes. It possesses 1 MG 34 on the right-hand side and, as additional armament, has 2 grenade launchers, about 80-81mm. It is believed to be powered by a Maybach engine. The vehicle had been driven 1687 kilometres according to the odometer. It is estimated that a 6-man crew is needed to operate the vehicle."[173]*

The next day, the US 30th Infantry Division, 7th Armored Division and US VII Corps all reported that a prisoner had been captured the previous day at Oberembt, and had belonged to Sturm-Mörser-Kompanie 1000. The US 1st Army also reported capturing a German document, which showed that Sturm-Mörser-Kompanie 1000 was attached to the 363.Volks-Grenadier-Division on 27 February.[174, 175, 176, 177]

26 February 1945. A super-sharp Signal Corps photo of the Sturm-Mörser in Neusser Straße, Oberembt, taken by T/3 Roger H. Zachary. The vehicle threw a track, and the crew debated whether or not to set demolition charges. They decided against this as it would have destroyed a large area. One of the crew tried to defend the position with a machine gun from the vehicle, but was killed by an American hand-grenade. His body can be seen under the straw at the front of the Sturm-Mörser. *NCA*

A British Sherman ARV Mk.I clatters past the massive beast, looking small in comparison. The Oberembt vehicle was not fitted with extra armour on the bow, and unusually, the front trackguards have 'Zimmerit'. The weapon was loaded when found. **TTM**

A companion shot to that on page 48, with the dead crewman in front of the vehicle. Note the damage to the first outer roadwheel which, unlike the other roadwheels, is not covered in mud. **NCA**

RODNA

Opposite and above: More of the front of the Oberembt vehicle. **Right**: The dead crewman has been removed. *3x TTM*

The Oberembt Sturm-Mörser generated quite a lot of interest with the Army and GIs alike. With two Americans clambering over the vehicle, this shot shows the MP-port hanging from its chain on the side of the fighting compartment. When fully loaded with ammunition, it is debatable how useful this would have been. **D.Brown**

We can only imagine how many GIs had their photos taken in the huge barrel. **US Army**

A clear front view, with US infantry and tankers. The track broke after entering or trying to back out of the drainage ditch. With the projectile removed from the barrel, we can see into the fighting compartment. *T.Haasler*

The American soldier with the long pole could be working on a communication line. Once again, the Sturm-Mörser has attracted a crowd. *T.Haasler*

US Technical Intelligence officers took these two photos with a flashgun in diminishing light. This makes the three-colour camouflage pattern much clearer. The photo on this page has the crane (1), radio operator's MG ball mount (2), and driver's vision arrangements (3) indicated. The driver's visor has only one periscope in place. *2x C.Lemons via D.Neely, NARA*

This low side-view shows the damaged first roadwheel and battered trackguard. The Tiger Ausf.E had several tools fitted on the hull top, which were relocated to the sides when converted to a Sturm-Mörser. We have not yet seen a photo of one of these vehicles with a complete set of tools in situ. *C.Lemons via D.Neely*

Pages 59-63: The back end. **This and next page:** A US tank destroyer hit the Sturm-Mörser in the engine compartment with four hits; three of them closely grouped. The photo also shows small arms or HE scars. **Page 62:** The hatch on the ground is the rear of the loading hatch with 'Nahverteidigungswaffe'. These rear photos show a detail not mentioned in other books: a fifth shot that ricocheted off the top of the engine compartment and hit the back of the superstructure. **Page 63:** This excellent view was taken from the upper story of the buildings behind and shows that one of the roof hatch hinges has broken. *2x TTM, 2x NCA*

Sturm-Mörser-Kompanie 1000
Oberembt, Germany
26 February 1945
RODNA

This page: A US Technical Intelligence officer shows the size of the barrel using his arm. The interior of the vehicle appears to be illuminated. Unfortunately, the left corner of the photo has been torn off. **Opposite:** The Raketenwerfer 61 at maximum depression: zero degrees, which was how it was loaded.
2x C.Lemons via D.Neely

This page: The gun at maximum elevation. The outer barrel of the mortar was a single piece casting, as can be seen here. **Opposite:** Taken from the roof of the fighting compartment, the massive casting of the gun mount and barrel are apparent. At the top right of the photo is the Sturm-Mörser's track.
1x C.Lemons via D.Neely, 1x NCA

This page: More of the Raketenwerfer 61, showing its cast texture and the loader's MG ball mount. **Opposite:** The rear of the fighting compartment with aerial base, crane, and opened roof hatch with 'Nahverteidigungswaffe'. The cylindrical object just inside the rear hatch is the counterweight for the opened roof hatch. A spare aerial was carried above the rear hatch. *2x NCA*

This page: The roof with opened hatch, one of its latches missing. To the right is the rotating periscope. The three spigots welded to the roof and the loading hatch were for a rangefinder. **Opposite:** The 'Nahverteidigungswaffe' in detail. It could fire smoke, flare and anti-personnel rounds. *2x NCA*

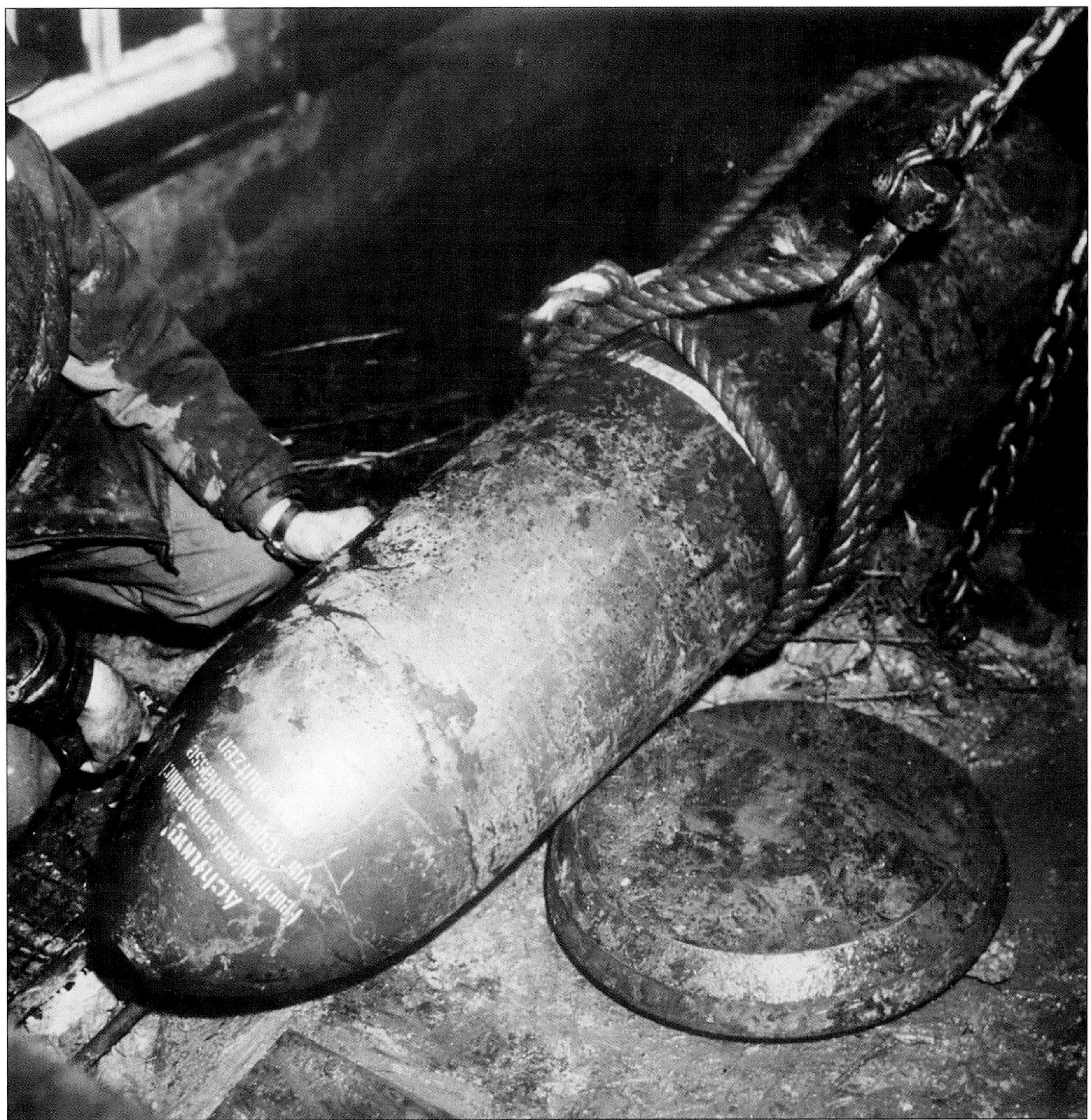

The ammunition is removed and inspected on the engine deck. **Opposite:** Cigarette in mouth, a US soldier measures the projectile with the help of a winch from a wrecker. **This page:** Another projectile is removed, this time in the rain. *1x TTM, 1x LAC*

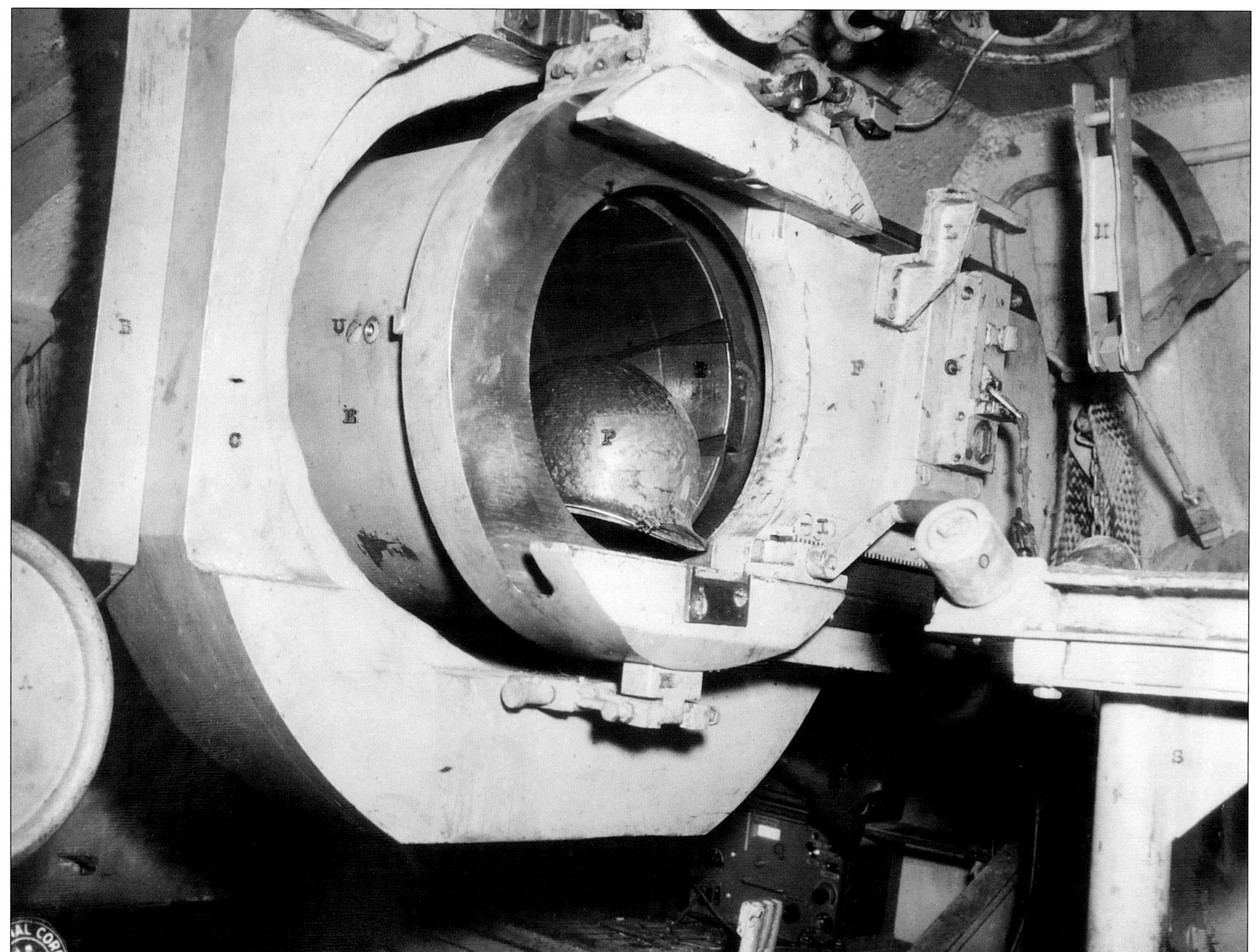

Inside the Oberembt Sturm-Mörser. **Opposite:** A US helmet is used as a size comparison inside the barrel. At the bottom of the photo, the radio can be seen as well as the radio operator's MG34. **This page:** The opened breech showing the rifling grooves inside the barrel. Splines on the projectile engaged with these grooves. *1x LAC, 1x NARA*

The detritus of war. Although not identified, it is likely that this photo is of the Oberembt vehicle. Three fuzes lie on the floor along with a bag of 7·92mm ammunition and a Jerry can. The Jerry can is sitting on one of the large cross members that ran from side to side. In the bottom right of the photo is the loading trolley. *K.Münch*

The left ammunition rack. Like the photo before, it is not identified as the Oberembt vehicle. The object on the left of the photo is part of the rotating periscope mechanism. ***K.Münch***

Opposite: Looking through the rear hatch at the loading trolley, which could be folded away when not in use. When loading the weapon, the projectiles were positioned so that one spline faced the floor, and to facilitate this, the top of the trolley was grooved. The large projectiles can be seen in their racks, while in the background is the headpad of the radio operator's MG34. ***NCA. This page***: The radio operator's position seen from below the weapon. His seat back has been raised - compare with the photo on page 30. The radio was fitted in a rack in the sponson; in the Tiger tank this area was used for MG ammunition. The brackets on the side wall are in the same position as those in the Tiger tank; a respirator and spare prism. The light coloured object in the right of the photo is the bottom of the loading trolley. ***C.Lemons via D.Neely***

Compared to the Tiger tank, the majority of the driver's controls were the same. The obvious differences are the new grab handle (behind the steering wheel) and relocated instrument panel. The back of the driver's seat has been removed and placed in the sponson. *C.Lemons via D.Neely*

The Oberembt vehicle was evaluated at the School of Tank Technology in Chertsey, UK. It is shown attached to a Bergepanther Ausf.A, chassis number 175644, but only by one tow cable. By this time, the Sturm-Mörser had gained several painted-on notes, among which is its designation under the gun. In the background is the hull of a Churchill tank. *TTM*

RECOVERED
By 464 ORD.
EVAC.
DTD
3107

The following series of photographs were taken at the School of Tank Technology, where it was allocated the number 3107 by the Department of Tank Design (DTD). These two front 3/4 views show flattened hooks and scratching to the side of the fighting compartment. Rather than move the camera around an immobile vehicle (remember it was knocked out by a series of hits into the engine compartment), the Sturm-Mörser has been moved for each photo - note the second roadwheel with its oil stain in different positions. ***1x L.Archer, 1x T.Haasler***

An imposing head-on view, and one that shows who recovered it: the US 464th Ordnance Evacuation Company, specifically Sergeants Yarmosh, Mabry and Johnson. The DTD 3107 serial number has been painted on both the hull and fighting compartment. Tiger tracks were not 'handed' as can be seen here; the same links being used for the left and right tracks. *TTM*

The scar from the ricocheted armour-piercing round to the left of the entrance hatch is visible here. Above this are the remains of what we believe are a pair of track hangers. The rear views taken at Oberembt show damage to one and the other missing. The AP round likely hit the tracks on these hangers, blowing one away and flattening the other. It did not survive the journey to the UK. ***TTM***

This page: A view of the fighting compartment roof minus the loading hatch. Other books have speculated about the use of the small circular armoured object next to the commander's vision hatch, but the reality is that its purpose is unknown. Most vehicles had a circular plate tack-welded inside the opening, as this example does. **Opposite:** With the fighting compartment removed, we can see how the Tiger hull was converted: by cutting away the hull roof, installing a new fighting compartment floor, and fitting large reinforcing bulkheads behind the driver and radio operator. This view clearly shows how little space the commander, loader and gunner had. At the top right of the photo are the empty radio racks. *2x TTM*

The superstructure has been removed. **The page:** The gunner's position showing the mounting for the gunsight, armoured shutter in the closed position, MP port and five opened 'Atemslauch' tubes. **Opposite:** This could be considered the loader's position. The gun mount allowed for 10° of traverse left and right, which explains the large gap to the side of the gun. The radio operator's machine gun mount (minus machine gun) is visible, as is the electrical conduit running vertically from the radio, then towards the aerial at the back of the vehicle. Note the crew light and switch for the ventilator. *2x TTM*

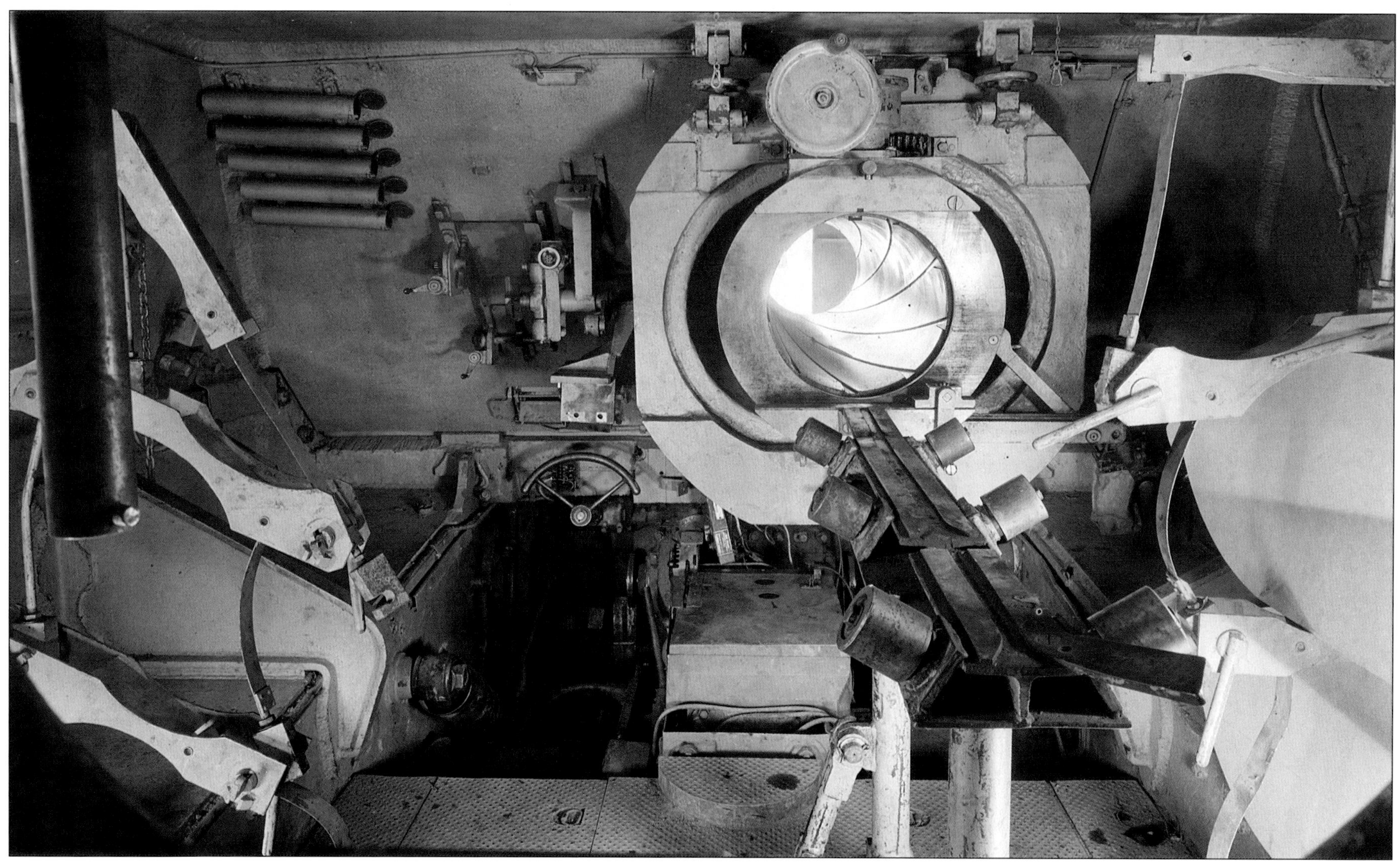

This page: Taken before those on the previous pages. With no ammunition, the fighting compartment seems quite spacious. Note the groove in the top of the loading trolley and how it aligns with the rifling inside the barrel. The gunner's elevation handwheel has been removed, as has the breech plate, see opposite. **Opposite top:** The outer face of the breech plate showing the toothed rack at the bottom and firing mechanism. **Opposite bottom**: The inner face of the breech plate showing the perforated ring. *3x TTM*

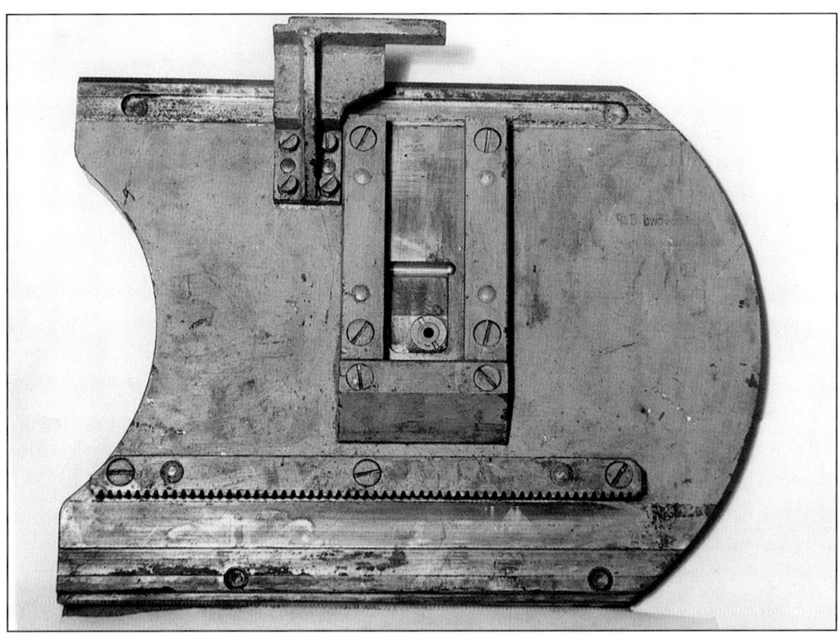

Except for a single combat readiness report, information about Sturm-Mörser-Kompanie 1001 for February 1945 is vague. It was said to have been committed along the Roer front with Sturm-Mörser-Kompanie 1000. After the Allies went on the offensive on 23 February, the company started to fall back to the East. Passing through Euskirchen, the company reached the Rhine near Bonn.[178] According to prisoners later captured in the Ruhr pocket, one Sturm-Mörser was destroyed by its crew during the retreat to the Rhine.[179]

In February 1945, Sturm-Mörser-Kompanie 1002 was finally released for combat and attached to the Oberbefehlshaber West on 2 February[180] with six Sturm-Mörsers. The commanding officer was Oberleutnant Rudolf Zippel.[181] On 11 February, the company was attached to the 15.Panzer-Grenadier-Division, which by that time was fighting British forces in Weeze and the Reichswald. The same day, a combat force of Panzerjäger-Abteilung 33, under the command of Oberleutnant Wiener moved to the area east of Wissen Castle, 1·5 kilometres southeast of Weeze, while the battalion command post was located in Triphof, 2 kilometres east. The force consisted of three Panzer IV/70(V) and two 7·5cm Paks. An element of Panzer-Abteilung 115, also from 15.Panzer-Grenadier-Division, was attached to the Panzer-Abteilung with one Befehlspanzer and one Sturmgeschütz III. Sturm-Mörser-Kompanie 1002[182] also supported the anti-tank unit with its five Sturm-Mörsers. At 19:30, the combat force was reinforced by two more Pak 40s from the battalion's 1st Company and moved to a new assembly area at Asperberg on the southeastern edge of the Reichswald.[183] On the night of 12/13 February, heavy enemy artillery fire fell on the German lines. Another anti-tank platoon with an unknown number of 7·5cm guns arrived, while a Panzer IV/70(V) bogged down, and because no recovery vehicles were available, was destroyed by enemy fire. A Panzer IV and Sturmgeschütz III arrived from Panzer-Abteilung 115. In the afternoon, the five Sturm-Mörsers from Sturm-Mörser-Kompanie 1002 opened fire on targets and assembly areas in the Reichswald. A total of forty rounds were fired. Based on the enemy's reaction, they had a significant effect.[184] The 15.Panzer-Grenadier-Division historian, Kurt Albert Rust, stated that Sturm-Mörser-Kompanie 1002 was attached to Panzer-Abteilung 115 and Panzerjäger-Abteilung 33 again on 17 February 1945. That day, the Sturm-Mörsers were reported as combat-ready. Their engagement did not last very long. Due to their heavy weight, several vehicles got bogged down in the fields and had to be blown up to avoid capture.[185] Regardless of whether this statement is true, the company was once more attached to the 15.Panzer-Grenadier-

Division on 24 February. This time only two Sturm-Mörsers were reported as available.

That day, enemy forces advanced along the Goch-Weeze road in the direction of Weeze. The 15.Panzer-Grenadier-Division tried to slow the advance with Panzerjäger-Abteilung 33, Heeres-Panzerjäger-Abteilung 741 (equipped with Hetzers), and one company of Panzer-Regiment 130 (Panzer-Lehr-Division) (equipped with Jagdpanthers). In the evening, the two operational Sturm-Mörsers of Sturm-Mörser-Kompanie 1002 were attached to Panzerjäger-Abteilung 33. The vehicles only had 12 projectiles between them. Both opened fire on identified enemy assembly areas.[186] From 26 February until 3 March, the company was attached to the 116. Panzer-Division West of Xanten.[187] On 28 February, while attached to the division and in support of I./Panzer-Regiment 16, it deployed two Sturm-Mörsers near Villa Reichswald.[188]

By the end of February 1945, the 16 Sturm-Mörsers were assigned as follows:

Stu.Mrs.Kp.1000	3	*(1 combat ready)*
Stu.Mrs.Kp.1001	3	*(3 combat ready)*
Stu.Mrs.Kp.1002	6	*(2 combat ready)*
Pz-Ers.- u. Ausb.Abt.500	1	*(prototype)*
HZa Sennelager	3	*(2 assigned to Stu.Mrs.Kp.1000)*

March 1945

No information has been found that shows what happened to Sturm-Mörser-Kompanie 1000 and its three remaining Sturm-Mörsers after crossing the Rhine in early March. It is highly likely that it crossed the Rhine at Cologne and moved to an assembly area near Bergisch Gladbach for rest and refitting.

In early March, Sturm-Mörser-Kompanie 1001 crossed the Rhine by ferry near Bonn. There is conflicting information about the number of Sturm-Mörsers that crossed. After the war, Heinz Matten claimed that only three vehicles arrived east of the river. Two other men from the company confirmed Matten's statement. The Americans captured Unteroffizier Hans Döring and Obergefreiter Heinrich Höfemeyer in the Ruhr pocket on 15 April 1945; both claimed that one Sturm-Mörser was destroyed by the crew west of the Rhine. Another soldier, a tank driver captured in May 1945, stated that all four vehicles had crossed the river in March. The company stayed east of Bonn until at least 21 March. While there, the last of the ammunition was fired on American positions.[189, 190, 191] The company appears to have been attached to the 353.Infanterie-Division, as it was mentioned in the distribution list of the *"Special instructions for supply"* issued on 14 and 21 March. The divisional supply and transport elements were in the area east of Siegburg-Troisdorf.[192, 193] By the end of March, the company moved to Odenthal near Bergisch Gladbach with just one projectile left. Sturm-Mörser-Kompanie 1000 was in the same area by this time. What happened next is again not clear due to conflicting statements, but at least one Sturm-Mörser was reissued to Sturm-Mörser-Kompanie 1000 to make up for the loss west of the Rhine in February.[194, 195, 196, 197]

The attachment of Sturm-Mörser-Kompanie 1002 to Kampfgruppe Tebbe of Panzer-Regiment 16 ended on 2 March, after the recapture of Hufscherberg. The company fired four projectiles on the area around Villa Reichswald, virtually destroying the 'Villa'. One of the rounds also hit an enemy ammunition dump, which exploded in a spectacular fashion.[198, 199]

After the action at Hufscherberg, the company crossed the Rhine at Rheinberg and was sent back to Paderborn for rest and refitting. They still had six Sturm-Mörsers, but only three were combat-ready, with two in short-term maintenance and one in long-term maintenance. While at Paderborn, Hauptmann Wilhelm von Gottberg took over command from Oberleutnant Zippel. The unit returned to the front at the end of March and was once again attached to the 116.Panzer-Division with four Sturm-Mörsers.[200, 201, 202] Whether the two missing Sturm-Mörsers stayed behind at Sennelager due to technical problems is not known.

Around 28 March, the company reached the Dorsten area, where it was attached to I./Panzer-Regiment 16, which was withdrawing east from the Wesel area via Kirchhellen. The Sturm-Mörser company's first engagement took place that day east of Kirchhellen. After the village had been abandoned by the Germans, Sturm-Mörser-Kompanie 1002 moved to a firing position in the woods north of Feldhausen, about 2 kilometres east of Kirchhellen. From this position, it had a clear view of the road junction north of Kirchhellen. A large group of American vehicles was identified in the area that was either forming up or held up. All four Sturm-Mörsers aimed directly at the target and fired a salvo. By the standards of the time, the impact was unbelievable. Vehicles flew through the air and tanks flipped over like cardboard boxes. Immediately after firing, the Sturm-Mörsers left their positions, but the Americans took hours to recover from the shock and sent up reconnaissance aircraft. These American forces in the Kirchhellen area were probably the 30th Infantry Division or 8th Armored Division. Hauptmann Adam, the commanding officer of I./Panzer-Regiment 16, thought the impact of the rockets was devastating. The rocket-propelled projectiles left a trail of smoke though, which revealed the Sturm-Mörsers' firing position, so, they had to move after firing every projectile.

Major Tebbe, the commanding officer of Panzer-Regiment 16, also remembered that a battery of Sturm-Mörsers was assigned to his regiment, while at the northern edge of the Ruhr. He committed the battery in a situation he considered tenuous. From his point of view, the impact of the fire was phenomenal. The enemy stopped his advance for quite some time after being hit. In general, everybody regretted that this weapon had not been not available before. Major Tebbe reported the commitment of the battery to his divisional headquarters and was surprised to hear that everyone was shocked to learn of its use. Tebbe had to justify his decision. The reason was that there had been an O.K.H. directive, which allowed the commitment of the battery only by its approval and that of the Führer headquarters. The approval request required a justification from a tactical perspective. Tebbe was glad to hear that the Division Commander received retrospective approval.[203, 204, 205, 206, 207]

By the end of March 1945, the 16 Sturm-Mörsers were assigned as follows:

Stu.Mrs.Kp.1000	4
Stu.Mrs.Kp.1001	2
Stu.Mrs.Kp.1002	4
Sennelager	2 *(assigned to Stu.Mrs.Kp.1000)*[208]
Sennelager	2 *(assigned to Stu.Mrs.Kp.1001)*
Sennelager	1
Unknown Location	1 *(prototype)*

April/May 1945. The Hützemert Sturm-Mörser

Sturm-Mörser-Kompanie 1000 was in the Bergisch Gladbach area until early April. To make up for the loss of the Sturm-Mörser at Oberembt, it received one vehicle from Sturm-Mörser-Kompanie 1001. Leutnant Karl Hubert Doll commanded this. After the war, he recalled that after the Allied breakout from the Remagen Bridgehead and the advance on Siegen, he was ordered to report to Hauptmann Franz Kodar, the company commander, on 9 April 1945. He received the simple order to move towards the enemy and support the troops in the Drolshagen area. After his vehicle was refuelled and resupplied with ammunition from the other Sturm-Mörsers, Doll and his crew started to move towards Drolshagen that evening. Under cover of darkness, they were at least safe from Allied aircraft. But moving along the unfamiliar, narrow and winding roads of the Bergisches Land without headlights was exhausting for the crew. To complicate the situation, most of the roads were clogged with other forces moving in various directions. In Marienheide, north of Gummersbach, they refuelled the Sturm-Mörser with 400 litres of petrol. East of Marienheide, an Army Oberst stopped the vehicle with a torch. He thought Doll and his crew were driving in the wrong direction, as the enemy was already nearby. Doll explained that he had orders to move to Hützemert to support troops in the Drolshagen area. The Oberst replied that there were no longer any German units that Doll could support. He also said that his men had erected roadblocks to slow the Americans. Doll explained that roadblocks were no obstacle for his vehicle unless they were mined. The Oberst confirmed that the roadblocks were not mined, and after Doll told him he had orders he intended to obey, the Oberst bid him and his crew farewell. The road became more and more congested with retreating German troops, so Doll decided to use a secondary road, which bypassed the roadblocks. Doll ordered a stop at Lake Agger to give the crew some rest. The driver in particular was exhausted after travelling almost 60 kilometres in total darkness. After two hours of sleep, they set off again and reached Hützemert at dawn. They went into position on the southeastern edge of the village, called 'Im Herrenscheid'. From here, they could see Drolshagen and the road running from Hützemert to Drolshagen.[209]

Before noon, Hauptmann Kodar arrived in Hützemert and inspected the position of the Sturm-Mörser. Kodar and Doll then went on reconnaissance in the company commander's vehicle. They drove through Drolshagen and headed for Olpe but were stopped by enemy artillery fire near Schmierhagen so they returned to Herrenscheid. Except for a single German artillery gun that tried to move to a new firing position near Schmierhagen, they had not seen any German forces, roadblocks or positions prepared to engage the Americans. Doll started to recall the words of the Oberst he met the night before.

Despite the disappointing reconnaissance, Hauptmann Kodar said goodbye without issuing any other orders. Several hours later, a motorcycle messenger arrived with orders from the company commander to fire one projectile at the steeple in Drolshagen. Doll was surprised and asked the messenger if the Americans were in the town. The soldier could not answer, which raised major concerns for Doll. Since Kodar left before noon, Doll had carefully watched the terrain and had not spotted any troop movements. Everything was quiet. There was no artillery fire and even the ever-present American reconnaissance aircraft were not flying. He believed that his company commander had misjudged the situation and shared his observation with the messenger. He told him to inform Hauptmann Kodar that he intended to fire on the American armoured spearhead as soon as it approached Drolshagen from Schmierhagen. Leutnant Doll hoped that Hauptmann Kodar would accept his recommendation and modify his order. A projectile fired at the steeple might have killed some Americans but also numerous civilians, not to mention the effect of the impact on the church and the surrounding buildings.

After the messenger left, Doll realised that he had just disobeyed the orders of his commanding officer. He talked to his crew to get their points of view. All agreed not to fire on Drolshagen, but surrendering to the Americans was also out of question. In case Hauptmann Kodar insisted on executing his order, the crew decided to move towards the enemy and to engage them in open terrain. Doll reminded his crew that their decision would be a clear refusal to obey orders, and that they would have to bear the consequences, but they did not change their minds. The messenger returned after 30 to 45 minutes and told Doll that Kodar was fuming, and demanded he open fire on the steeple immediately. Everyone was upset, as they hoped that Kodar would change his mind. Doll told the messenger that they had made their decision and refused to fire on the steeple. He dismissed the messenger with the words: *"Leutnant Doll and his crew are signing off."*

After the messenger left, the crew mounted their Sturm-Mörser and started to move in the supposed direction of the enemy. It was clear that they could not stay in Herrenscheid because they assumed that Hauptmann Kodar would return soon with another crew that would carry out the order. Their plan came to a sudden end, when the left track seized with a loud noise,[210] while they still were in Hützemert near the main road to Drolshagen. The vehicle was immobilised, and the weapon could no longer be aimed. In other words, the Sturm-Mörser was not combat-ready. Under the circumstances, it was impossible for the crew to repair the vehicle on their own. The vehicle had stopped between a house to the right and a bank to the left; it was at least concealed from the direction the crew expected the enemy to advance.[211]

Up to that point, no enemy activity had been identified. After some time, enemy observation aircraft were seen, and sporadic artillery fire fell on Hützemert. Doll once more asked his crew, if they should surrender in case the Americans arrived. Again, his men refused. They also considered blowing up the vehicle, but with high-explosive projectiles still inside, this would have had the same effect on Hützemert that they had just prevented in Drolshagen. This was not an option for the crew either. A short while later, the first American tanks entered the village but initially failed to discover the Sturm-Mörser. The crew left the vehicle and prepared to fight off any Americans that might come in their direction. When the Americans finally discovered the Germans, there was a firefight with small arms that lasted about half an hour.

When the fighting ended, Unteroffizier Werner Herrler, the gunner, and Gefreiter Karl-Heinz Langer, the loader, had been killed, while Gefreiter Gerhard Gäbler, the radio operator, and Leutnant Doll were wounded. Doll managed to hide in a nearby house, where he met a civilian doctor from Bonn, who dressed his wounds. When Doll tried to leave the house to continue the fight, the civilians in the house persuaded him to surrender instead. He handed his pistol to them and was soon captured by the Americans, most likely from 1st Battalion, 13th Infantry Regiment (8th Infantry Division), who took him to a military hospital in Drolshagen. On their way to Drolshagen, they discovered the seriously wounded radio operator, who was lying on the road, and took him with them. Despite an emergency operation, Gäbler died from his wounds. The fate of Doll's tank driver remained unclear. The three men killed of his crew were buried in Drolshagen cemetery.[212, 213, 214]

On 12 April 1945, the 1st US Army reported that fourteen men from Sturm-

Mörser-Kompanie 1000 had been captured, but no place or capturing unit was included.[215] The next day, Company C of the US 893rd Tank Destroyer Battalion claimed to have destroyed two 38cm self-propelled rocket guns. Again, no place was given. The company was attached to the 311th Infantry Regiment (78th Infantry Division) between 1 and 17 April. It is known that the regiment was in action in the Marienheide area north of Gummersbach on 12 April.[216] It is an educated guess that these two Sturm-Mörsers belonged to the remnants of Sturm-Mörser-Kompanie 1000, which were out of fuel and ammunition.

It was not until 13 April that the Americans reported the capture of the Sturm-Mörser near Drolshagen. Reports from the US 1st Army mentioned that it was captured almost intact and that it belonged to Sturm-Mörser-Kompanie 1000, although there was differing information regarding the number of captured soldiers. One report mentioned one captured soldier, most likely Leutnant Doll, while another report claimed that the unit had been used as infantry and the majority surrendered.[217] The second report is ambiguous, if the information was referring to the capture of soldiers in the Drolshagen area or if it was referring to the capture of several soldiers on 12 April. Over the next few days, more men from Sturm-Mörser-Kompanie 1000 were captured in the Ruhr Pocket: three to four on 17 April, four on 18 April and two on 22 April.[218, 219, 220, 221]

In early April, Sturm-Mörser-Kompanie 1001 detached one Sturm-Mörser from Odenthal to the area of Cologne in the expectation of an American attack. Lacking fuel and ammunition, Oberleutnant von Stapff, the commanding officer, finally issued orders for the crews to destroy their vehicles. The company was disbanded, and the personnel attached to the 353.Infanterie-Division. One of the company's officers was captured in the Ruhr Pocket on 16/17 April 1945. Like so often at the end if the war, no location was given.[222, 223, 224, 225, 226]

From photographic evidence, we know that at least five different Sturm-Mörsers managed to escape the Ruhr Pocket in April 1945. It is not known if these were the remains of Sturm-Mörser-Kompanie 1002, which had withdrawn with the trains of 116.Panzer-Division via Beckum and Hameln towards the Harz region,[227] or if these were the vehicles that had been at Sennelager at the end of March.[228] The path of three Sturm-Mörsers can be traced during the first week of April 1945. On 1 April, three Sturm-Mörsers stopped in Rietberg on Mastholter Straße between the southern gate

and the junction to Bokeler Straße. Under pressure from the advancing Americans, the 20 soldiers and 3 vehicles withdrew in the direction of Delbrück before noon.[229] On 6 April, civilians reported seeing three Tiger tank chassis armed with 380mm mortars, camouflaged in Hameln on the afternoon of 4 April.[230] On 5/6 April 1945, the elements of 116.Panzer-Division that had escaped the Ruhr Pocket reached the western edge of the Harz Mountains near Clausthal-Zellerfeld. On 7 April, the staff received orders from Oberbefehlshaber West, to immediately raise a Kampfgruppe from the remnants of the 116.Panzer-Division. It was to have the following order of battle:

- Headquarters with a mixed signals company
- One tank company with 14 Panzer V's
- Two armoured infantry companies, each with 8 medium armoured personnel carriers, with 80-100 enlisted men per company
- One mixed company with one antiaircraft platoon, consisting of 5x 2cm guns (towed), 3x 3·7cm guns (towed) and 2x 8·8cm guns (towed); one reconnaissance platoon, consisting of three sections (two sections with 2 light scout cars each and one section with two 8-wheeled scout cars; one platoon with mortars on Tiger chassis (4 mortars)
- One supply company
- One motor transport company 50 tons[231]

Major Brühl became the commanding officer of the Kampfgruppe. He had been the commanding officer of II./Panzer-Regiment 16 until he fell ill in December 1944. The history of the 116.Panzer-Division states that the Kampfgruppe never reached strength, because insufficient men escaped the Ruhr Pocket.[232] It is also very unlikely that the four vehicles were attached to the Kampfgruppe. It looks as if they moved in an easterly direction north of the Harz Mountains. The Sturm-Mörsers probably tried to reach Burg near Magdeburg, the home of the German assault gun artillery, which had supplied replacements since Sturm-Mörsers were assigned to the artillery in January. This assumption is based on the time and location of two vehicles captured by the Americans near Magdeburg. It is also possible that the Sturm-Mörsers received orders to move to Berlin for the defence of the German capital. A third Sturm-Mörser was captured at Rogäsen, 20 kilometres southwest of Brandenburg and about 45 kilometres east of Burg, by the Soviet 3rd Army in the first week of May.

Here is the best-known photo of the Hützemert Sturm-Mörser, taken by T/6 R.E. Michaud on 11 April 1945. The location is Schulweg, just off the junction with Hauptstraße. The vehicle broke down and was abandoned by the crew. Spare tracks were carried on the sides of most Sturm-Mörsers, in the same manner as the Panther tank. The original photo caption claims the vehicle had a crew of nine. **NARA**

RODNA

A photo from M/Sgt William H. Hunter of the US 744th Light Tank Battalion showing a soldier preparing to cut the tracks with an oxyacetylene torch. The reason for their requiring cutting is evident here: the Sturm-Mörser's drive sprocket has disengaged from the track, making it impossible to move. Cutting the track at the front and rear allowed the roadwheels to rotate freely and enable a straightforward recovery. The 744th Light Tank Battalion occupied Olpe in April 1945, about 9km from Hützemert. The vehicle had to be removed because it was blocking the road, as can be seen here. **R.Hunter**

In a letter to his family in Oklahoma, M/Sgt. Hunter wrote that the Sturm-Mörser still had five rounds of ammunition inside.

This photo is damaged but clearly shows the vehicle's proximity to the building on the left and exposed drive sprocket. This and the following three photographs were taken by Leo Becnel, also of the US 744th Light Tank Battalion. The photos of the right track coming off the sprocket are at odds with the testimony of the crew, who stated that the left track seized. *L.Becnel via G.Boyd*

To our knowledge, the left side of the Hützemert Sturm-Mörser has never been seen in print. Note the curled up tow cable on the side and rudimentary lifting ring on the loading hatch. *2x L.Becnel via G.Boyd*

The Sturm-Mörser was towed to a railway yard, probably in Siegen, about 35km from Hützemert. Not having a set of tracks has increased its ground pressure, causing the 68-ton vehicle to sink into the ground. The hooks on the side of the fighting compartment are thinner and flimsier than those seen on other Sturm-Mörsers. Note that the sprocket ring has been removed. *L.Becnel via G.Boyd*

The remainder of the tracks are on the engine deck and in the background, along with a Panzer IV/70(V), presumably from 116.Panzer-Division, who were in the area. To the right, the gun barrel from a Hummel pokes into view. *AMC*

The Hummel, whose barrel is seen on the previous page. The projectiles are not from the Hummel or Sturm-Mörser (which is on the right of the photo). To the left is a Sd.Kfz.7 'Pritschbau'. A m.S.P.W. (Sd. Kfz.251) is visible behind the Hummel. **L.Becnel via G.Boyd**

An overview of the railway yard probably at Siegen, complete with two Sd.Kfz.7 'Pritschbau' and a 10·5cm leFH in the foreground. *L.Becnel via G.Boyd*

The Brumby Sturm-Mörser

Werner Schacke from Brumby near Calbe remembered that a Sturm-Mörser broke down on 11 April on Staßfurter Weg. Five soldiers got out and asked him where they could get fuel. There was no fuel available, so an NCO fired a Panzerfaust at it, which damaged the interior. The Sturm-Mörser, from Stu.Mrs.Kp.1002, was later examined by the Americans and remained in place for several months before it was destroyed. What remained was later scrapped by the steel plant at Calbe.[233]

Previous page, this page and opposite: The Brumby Sturm-Mörser was destroyed by the crew on 11 April after running out of fuel. The wreck became a popular attraction for passing GIs before being scrapped.
2x K.Münch, 1x L.Archer

This page and opposite: Men the US 83rd Infantry Division check out the wreck. Note the battle scar under the driver's visor and that the rifled liner from the barrel has been blown out. The rear view opposite shows significant blackening around the engine compartment and that the superstructure has been dislodged. The Sturm-Mörser has 'Zimmerit' on the bow and rear plates, but not the hull sides. *2x NARA*

Another rear view, showing the missing entrance hatch, and three spare track links on the back of the fighting compartment. This Sturm-Mörser was not fitted with spare track brackets on either side of the engine compartment. The caps and heat shields are missing from the exhausts. *AMC*

This close up shows that the fighting compartment has parted company from the hull, blowing out the fixings. *P.Johnson*

This photograph was probably the first taken of the Brumby Sturm-Mörser, as the barrel liner has been blown out, which GIs are now investigating. Next to this is the roof hatch. In front of the crouching American on top of the vehicle is a simple sighting vane, added in the field; a detail seen on a number of Sturm-Mörsers. **TTM**

A rare image indeed. Wartime colour photos are rare; but this is the only known colour photo of a Sturm-Mörser. It was taken by Major Thomas A. Ligon of the 967th Field Artillery Battalion. The 967th were firing on Zerbst between 22 and 24 April, so the photo was probably taken then. On the ground in front of the vehicle is the crane and fillet of armour missing from page 113. **T.Ligon via D.Neely**

The Ebendorf Sturm-Mörser

Günter Schnick from Magdeburg had been evacuated with his family to Ebendorf. On 17 April, he and his father followed the Americans on their bicycles from some distance. The Americans were advancing from Ebendorf in the direction of Magdeburg-Neustadt. After they crossed the bridge over the highway southeast of Ebendorf, they discovered a German vehicle with a very short gun on the left side of the embankment, which had lost a track.[234] It is worth mentioning that Ebendorf had already been captured by elements of the US 30th Infantry Division on 12 April. It is probable that the Sturm-Mörser was already damaged on 12 April, or the day before.

The Americans initially towed the Ebendorf Sturm-Mörser to the German Army Proving Grounds at Hillersleben and later shipped to the United States before returning to Germany. It is now displayed at the Panzer Museum in Munster. The vehicle has the chassis number 251174 and was from a batch of Sturm-Mörsers manufactured by Alkett in September 1944. In October 1944, it was assigned to Panzer-Ersatz- und Ausbildungs-Abteilung 500 and finally issued to Sturm-Mörser-Kompanie 1002 in December.[235]

The Sturm-Mörsers captured at Brumby, Ebendorf and Menden all exhibit the same camouflage pattern, which leads to the assumption that the Brumby vehicle also belonged to Sturm-Mörser-Kompanie 1002.

Opposite and this page: The Ebendorf Sturm-Mörser was found next to a highway bridge in April 1945. Although the circumstances surrounding its abandonment are unknown, it is likely that the poor quality photo on this page was taken when found, and its track broken. It is logical to conclude that the photo opposite was taken after being pushed off the road. Its track was dragged from the road too, which might explain the track's snake-like appearance. We have no idea what the other cast steel items are. *1x NARA via D.Neely, 1x AMC*

Photos from Lt. Col. Lonsdale P. MacFarland of the US 5th Armored Division. **This page**: A GI peers into the entrance hatch. The vehicle has a 'fresh' look, with little weathering to the paintwork. Note the dark colour of the loading crane. **Opposite:** An imposing view of the Ebendorf Sturm-Mörser. *NWW2M 2009.131.120/2009.131.094*

The next five images were published in Panzerwrecks 2 and show the Ebendorf Sturm-Mörser in some detail. These two well-exposed shots clearly show the 'fresh' look to the paint scheme and the details on the roof. *2x J.Scott*

A similar photo to the previous page, but now with a GI. While over-exposed, the photo highlights the 'Zimmerit' pattern on the hull and cast texture of the gun mantlet. *J.Scott*

The same GI as the previous page posing with the Raketenwerfer 61. We can see straight through the barrel out to the landscape beyond. This Sturm-Mörser had the chassis number 251174 and was assembled by Alkett in September 1944. *2x J.Scott*

The Sturm-Mörser, complete with an American soldier poking out of the barrel. The photo was taken by Arthur D. Ballou of the US 557th Field Artillery Battalion. According to the markings on the front plate, it was to be transported to Depot 0-644, outside Paris. It was subsequently transported to the USA and later on, back to Germany. **D.Ballou**

RODNA

Preparing to move the 68-ton colossus onto a trailer. The vehicle has tow bars attached to the front tow points. The MP port, visible dangling on its chain in the previous photos, has been lost. This photo was presumably taken at Depot 0-644. **P.Kocsis**

The front view over the trailer reveals that it is a captured German Sd. Anh.121, which came in two capacities: 60t and 75t. *1x P.Kocsis, 2x D.Brown*

Here is the Sturm-Mörser ready to make the journey from Europe to the US. The missing track has been replaced and the Raketenwerfer barrel covered with a tarp. The sailor at the top was the owner of the photo, he was in the US Merchant Navy, and sailed between the US and the ETO several times during the war. *J.Norian*

An interesting photo taken upon its arrival at Aberdeen Proving Ground in Maryland. In addition to the shipping address painted on the side, is an inscription saying that the battery has been disconnected, it is free from fuel, and something else unreadable. Note how the old Tiger tank camouflage is visible below the missing trackguards and that the last roadwheel differs slightly from the others. **TTM**

While at Aberdeen Proving Ground, the Sturm-Mörser was given a fresh coat of paint and a 'Balkenkreuz'. **TTM**

Compared to other repaints of the era, this one is quite authentic, even if there are twice the number of 'highlights' on the darker areas. **USAHEC**

The vehicle was returned to Germany and is now exhibited at the Panzermuseum in Munster. This photo was taken in Germany and shows the trackless Sturm-Mörser being towed from a railway flatcar by an M60 tank. **USAHEC**

The Menden Sturm-Mörser

While withdrawing through Polsum, Marl and Datteln, Sturm-Mörser-Kompanie 1002 was said to have fired its remaining 24 projectiles from well-chosen positions. Hauptmann von Gottberg had to observe the requirements for his vehicles, which had to be coordinated with 116.Panzer-Division before every action. All of the Sturm-Mörsers were apparently blown-up east of Datteln after firing their last round. Out of fuel and with no chance of receiving ammunition, this sounds logical,[236] but at least one made it to the Menden area, where it was captured almost intact by the Americans. The Americans captured another vehicle, which had been assigned to the company near Magdeburg. Another source states that the remnants of the company were captured near Minden in April, but there is no evidence of this.[237]

On 14 April 1945, elements of the US 737th Tank Battalion (5 Sherman tanks) and supporting infantry of the US 2nd Infantry Regiment (US 5th Infantry Division) were advancing on Menden from the east, when a projectile exploded next to the lead tank, leaving a tank-sized crater. The infantry took cover behind the tanks, while the tanks tried to outflank the German gun position, with at least one American soldier being wounded. They soon captured a Sturm-Mörser and its crew at Oesberner Weg next to the Rauherfeld Quarry. The German crew were wearing black uniforms and initially thought to be SS soldiers. The Americans interrogated one of the prisoners, who claimed that the vehicle belonged to Sturm-Mörser-Kompanie 1002. Further investigation revealed that it was partially destroyed, but still had six 38cm projectiles nearby. Two days later, the US III Corps reported capturing members of Sturm-Mörser-Kompanie 1002 in the Ruhr Pocket.[238, 239, 240]

Another Sturm-Mörser; another GI in the barrel. The vehicle was captured on Oesberner Weg, next to Rauherfeld Quarry in April 1945. **K.Münch**

It looks as if the crew set a demolition charge in the engine compartment. The rear torsion bars have sagged, leaving the idler wheel almost touching the ground and the engine deck displaced. Like some other Sturm-Mörsers in this book, it has a rudimentary sighting vane welded onto the top of the glacis plate. *K.Münch*

RODNA

Unfortunately, the photos in this series are not of the highest quality. This view shows the nose up attitude caused by the explosion in the engine compartment and, like the other Sturm-Mörsers in this book, no markings. ***K.Münch***

A GI checks out the interior. The exhausts are missing their armoured outlets, caps and heat shields. The evergreen branches were probably on the vehicle as camouflage. At this point of the war, the Allies had air superiority, and this slab-sided monster would have been a sitting duck. **K.Münch**

Some of the foliage remains on the left side. Like other Sturm-Mörsers, spare tracks were fitted on the rear of the fighting compartment. *K.Münch*

Looking in through the rear hatch at the gunner's position. The gunner's elevation handwheel is of a spoked design, unlike the Oberembt vehicle seen earlier. The left ammunition rack is empty. **K.Münch**

With the breech open, the rifling is evident, as is the widening of the grooves to allow splines on the projectile to engage easier. The loading trolley is at the bottom of the photo. **K.Münch**

Two GIs try the barrel for size. The photo caption reads: *"Two of the boys in a big knocked out German tank in the Ruhr Pocket in April."* **L.Archer**

"Kenahan from Boston, and the knocked out German tank in Menden in the Ruhr Pocket late April 1945. Germany". The casting number can be seen on the gun mount, as can the sighting vane atop the glacis plate. ***L.Archer***

While it looks as if the Sturm-Mörser has been moved, it has not. The photographer is standing on the embankment seen in the background of the previous pages. *NWW2M 2013.606.162*

Some of the foliage even remains on the roof, under the tarp. We do not know the fate of this Sturm-Mörser, but like many AFVs, it will have ultimately ended up in the smelting pot. This and the opposite photo were taken by PFC John Diehl of the US 5th Infantry Division. ***NWW2M 2013.606.161***

The Rogäsen Sturm-Mörser

What happened to the third Sturm-Mörser that escaped through the Rietberg Gap on 1 April? We don't know for sure and can only speculate. As previously mentioned, the third Sturm-Mörser (the prototype) was probably the one captured by the Russians at Rogäsen. It was first moved to Kirchmöser (the home of Brandenburger Eisenwerke), tested at Kubinka and is now preserved at the Patriot Park in Russia.

The Americans captured two more Sturm-Mörsers outside the Ruhr pocket. In his book about the last days of World War II in the Wunsiedel area, Werner Bergmann mentioned a Kampfgruppe from Panzer-Regiment 3 (2.Panzer-Division), under the command of Oberleutnant Peter Prien, that withdrew from Kornbach (15 April) via Grub (18 April), Röslau and Bergnersreuth (20 April) to Waldsassen (23 April). The Kampfgruppe comprised 19 armoured vehicles:

- 2 Panthers
- 3 Panzer IVs
- 5 Hetzers
- 4-5 captured M3 armoured infantry carriers
- 1 Sturmtiger - 38cm rocket launcher
- 18 trucks[241]

Nothing more is known about the fate of the Kampfgruppe and what happened to the Sturm-Mörser. Without photographic proof, it is impossible to determine if this Sturm-Mörser is the vehicle photographed at Lichtentanne in Thuringia or the vehicle abandoned at Frontenhausen on 3 or 4 May 1945. Considering Lichtentanne is almost 120 kilometres northwest of Waldsassen, in the direction of the enemy, it is unlikely that this Sturm-Mörser is the one that was attached to Panzer-Regiment 3. On the other hand, given the timeframe - between 23 April and 3 May 1945 - and the distance between Frontenhausen and Waldsassen, about 210 kilometres, this Sturm-Mörser may be the same one that reached Waldsassen 10 days earlier. The area in question had not been fully taken by the Allies.

These three poor quality but rare photos show the prototype Sturm-Mörser waiting to cross the Rhine in March 1945. After its return from Warsaw, the prototype was given an armoured superstructure, steel wheels and a counterweight on the gun. Although the gun counterweight looks at first glance to be the same as the Hützemert vehicle, there are subtle differences. The radio operator's MG ballmount is of a different pattern. *3x K.Münch*

The prototype Stum-Mörser was captured by Russian forces in Rogäsen, where we see it. The vehicle was subsequently moved to Brandenburger Eisenwerke, then to Russia, where it is preserved today. Note the missing bolt-head from the extra armour on the bow (present in the previous photos) and what we assume is a depression limiter below the gun. *TsAMO*

Sturm-Mörser Kompanie 1001
Rogäsen, Germany
April 1945
RODNA

The Rogäsen vehicle during evaluation at Kubinka, Russia in 1946. The Russians have painted the thickness of the amour on some surfaces. The gun on the prototype had a lot more rifling grooves than production vehicles, which can be seen here. *V.Kulikov*

The Frontenhausen Sturm-Mörser

Around 30 April, before the arrival of American forces on 2 May 1945, a small Wehrmacht column with a Sturm-Mörser at the front drove, towards Frontenhausen from the direction of Dingolfing. Behind the Sturm-Mörser were two Opel-Blitz and two smaller trucks. The commander, a young Leutnant, had orders to take up position at Gangkofener Berg and fire on the approaching Americans. However, the wooden bridge over the Vils collapsed with the Sturm-Mörser on it. It was impossible to move forward or backwards. The crew tried to blow off the concrete base with explosives so that the vehicle could have freed itself, but this failed.

The other vehicles of the column stopped in front of the bridge; their crews then took off via Aiglkofen. There was no ammunition in the trucks; it was stored in the tank and later seized by the Americans. The trucks' contents were mostly plundered typewriters, accordions and bales of cloth, but hardly any food. The Police later returned these items.

How and why this unit reached the Vilsbrücke in Frontenhausen is unclear. One reason was undoubtedly the retreat of many German units to the south of the Reich. Another reason may have been the railway bridge at Marklkofen being blown up. This made onward transport from Dingolfing by rail impossible. A third reason could be that the commander of the Sturm-Mörser was related to Lieserl Wimmer. She and Leni Wöcherl hid the soldiers until after the Americans arrived and gave them civilian clothes. What subsequently happened to the soldiers is not known. There are said to have been no arrests after the arrival of the 318th Infantry Regiment. The Americans moved the Sturm-Mörser upstream with three successive blasts to clear the bridge, which they then repaired. The Sturm-Mörser remained in the river until the mid-1950s before being dismantled and transported away by Moll, a scrap dealer from Munich.[242]

The Frontenhausen Sturm-Mörser stranded in the river Vils. The end of the gun barrel has no lugs or counterweights The front of the superstructure has some additions in the form of camouflage loops, a battered sighting vane and rain guard over the gun mount. *R.Röbel*

Two of the large boltheads that fix the superstructure to the hull are missing. The bridge in the background was built by American engineers. *R.Röbel*

Inquisitive locals placed the wooden pole here so they could explore the wreck. *R.Röbel*

Local children pose with the Sturm-Mörser. The superstructure has been cut away, leaving the 150mm thick glacis plate and gun for a later date. In the left photo, we can see the engine access hatch is open. *2x R.Röbel*

The Lichtentanne Sturm-Mörser

Nothing is known about the Sturm-Mörser found in Lichtentanne. We have only found one photo this photo showing it abandoned in a wooded area after the war. It had no trackguards but a full complement of tie-downs on the glacis plate.

Unknown Unit
Lichtentanne, Germany
1945
RODNA

Allocation

Assignments and combat readiness according to Lage West, Lage Frankreich and various strength reports by the Inspector General of Armoured Forces.

Date	Stu.Mrs.Kp.1000		Stu.Mrs.Kp.1001		Stu.Mrs.Kp.1002	
	Attached to	Combat Ready	Attached to	Combat Ready	Attached to	Combat Ready
05/11/44		4		4		
10/11/44		4		4		
25/11/44		4		4		
10/12/44		4		4		
15/12/44	LXVII. A.K.		LXVII. A.K.			
16/12/44	LXVII. A.K.		LXVII. A.K.			
17/12/44						
18/12/44	LXVII. A.K.		LXVII. A.K.			
19/12/44						
20/12/44						
21/12/44						
22/12/44		3		4		
23/12/44						
24/12/44		3		4		
25/12/44		3		4		
26/12/44		3		4		
27/12/44		4		4		
28/12/44	OB West	4	OB West	4		
29/12/44	OB West	4	OB West	4		
30/12/44	OB West	4	OB West	4		
31/12/44	OB West	4	OB West	4		

Date	Stu.Mrs.Kp.1000		Stu.Mrs.Kp.1001		Stu.Mrs.Kp.1002	
	Attached to	Combat Ready	Attached to	Combat Ready	Attached to	Combat Ready
01/01/45						
02/01/45	OB West	4	OB West	4		
03/01/45	OB West	4	OB West	4		
04/01/45	OB West	4	OB West	4		
05/01/45	OB West	4	OB West	4		
06/01/45	OB West	4	OB West	4		
07/01/45						
08/01/45						
09/01/45	OB West	4	OB West	4		
10/01/45	OB West	4	OB West	4		4
11/01/45						
12/01/45						
13/01/45						
14/01/45	OB West	4	OB West	4		
15/01/45	OB West	4	OB West	4		
16/01/45						
17/01/45	OB West	4	OB West	4		
18/01/45	OB West	4	OB West	4		
19/01/45	Pz.AOK 6	4	Pz.AOK 6	4		
20/01/45	Pz.AOK 6	4	Pz.AOK 6	3		
21/01/45	Pz.AOK 6	4	Pz.AOK 6	3		
22/01/45	Pz.AOK 6	4	Pz.AOK 6	3		
23/01/45	Pz.AOK 5		Pz.AOK 5			
24/01/45	LXXIV. A.K.		LXXIV. A.K.			
25/01/45	LXXIV. A.K.		LXXIV. A.K.			
26/01/45	LXXIV. A.K.		LXXIV. A.K.			
27/01/45	LXXIV. A.K.		LXXIV. A.K.			

Date	Stu.Mrs.Kp.1000		Stu.Mrs.Kp.1001		Stu.Mrs.Kp.1002	
	Attached to	Combat Ready	Attached to	Combat Ready	Attached to	Combat Ready
28/01/45	LXXIV. A.K.		LXXIV. A.K.			
29/01/45	LXXIV. A.K.		LXXIV. A.K.			
30/01/45	LXXIV. A.K.		LXXIV. A.K.			
31/01/45	LXXIV. A.K.		LXXIV. A.K.			
01/02/45	LXXIV. A.K.		LXXIV. A.K.			
02/02/45						
03/02/45						
04/02/45	LXXIV. A.K.		LXXIV. A.K.			
05/02/45		4		3		
06/02/45	LXXIV. A.K.		LXXIV. A.K.			
07/02/45						
08/02/45						
09/02/45	LXXIV. A.K.		LXXIV. A.K.			
10/02/45	LXXIV. A.K.		LXXIV. A.K.			
11/02/45	LXXIV. A.K.		LXXIV. A.K.			
12/02/45						
13/02/45						
14/02/45	LXXIV. A.K.		LXXIV. A.K.			
15/02/45						
16/02/45	LXXIV. A.K.		LXXIV. A.K.			
17/02/45	LXXIV. A.K.		LXXIV. A.K.			
18/02/45	LXXIV. A.K.		LXXIV. A.K.			
19/02/45	LXXIV. A.K.		LXXIV. A.K.			
20/02/45	LXXIV. A.K.		LXXIV. A.K.			
21/02/45	LXXIV. A.K.		LXXIV. A.K.		XXXXIX. Pz.K.	
22/02/45						
23/02/45	LXXIV. A.K.		LXXIV. A.K.		XXXXIX. Pz.K.	

Date	Stu.Mrs.Kp.1000		Stu.Mrs.Kp.1001		Stu.Mrs.Kp.1002	
	Attached to	Combat Ready	Attached to	Combat Ready	Attached to	Combat Ready
24/02/45	LXXIV. A.K.		LXXIV. A.K.			
25/02/45	LXXIV. A.K.		LXXIV. A.K.			
26/02/45	LXXIV. A.K.		LXXIV. A.K.		XXXXIX. Pz.K.	
27/02/45	LXXIV. A.K.		LXXIV. A.K.			

Command and Leadership

Sturm-Mörser-Kompanie 1000

Hauptmann Franz Kodar
Kompaniechef
1 October 1944 - 16 February 1945 [243]

Leutnant Henner von Rennenkampff
Zugführer
1 December 1944 - 16 February 1945

Sturm-Mörser-Kompanie 1001

Oberleutnant Hans-Joachim von Stapff
Kompaniechef
1 October 1994 - 14 February 1945

Leutnant Hans-Wolfram Schmidt
Zugführer I
1 October 1944 - 27 December 1944

Leutnant Wolf-Dieter Schmidt
Zugführer
As of 14 February 1945

Leutnant d.R. Adolf Neuper
Vorgeschobener Beobachter
As of 14 February 1945

Leutnant Karl Hubert Doll
Zugführer

Sturm-Mörser-Kompanie 1002

Oberleutnant d.R. Rudolf Zippel
Kompaniechef
As of 26 February 1945

Hauptmann Wilhelm von Gottberg
Kompaniechef

Leutnant d.R. Günter Peschke/Paschke
Zugführer
As of 26 February 1945

Leutnant d.R. Günter Reutlinger
Zugführer
As of 26 February 1945

Leutnant Wolf-Dietrich Schmidt
Zugführer
Effective 29 March 1945

Sources:
NARA, T78, R946, Officer reports 14 February 1945 for Stu.Mrs.Kp. 1000, 16 February 1945 for Stu.Mrs.Kp. 1001 and 26 February 1945 for Stu.Mrs.Kp. 1002.
Doll: Kriegszeit und Kriegsende im Drolshagener Land
Guderian: 116.Panzer-Division

Kriegsstärkenachweisung Nr. 1161 15/09/44 Panzer-Sturmmörserkompanie "Tiger" (Panz.Stu.Mörs.Kp. "Tiger")	Head count			Weapons				Vehicles			
	Officers	NCOs	Enlisted Personnel	Rifles, Carbines	Pistols (Machine Pistols)	Heavy MGs (Light MGs)	Limbered (Unlimbered) Guns and Launchers	Limbered (Unlimbered) Vehicles	Motorcycles (MC Sidecars)	Motor Transport (Lorries)	Prime Movers (Fully Tracked Prime Movers)
A. Command Group *Gruppe Führer*											
Company Commander	1	.	.	.	1	.	.	.	.	.	.
Forward Observer	1	.	.	.	(1)	.	.	.	.	.	.
Ammunition Officer *(this entry was pencilled in)*	1	.	.	.	1	.	.	.	.	.	.
Company Headquarters Section *Kompanietrupp*											
Headquarters Section Leader *(also maintains daily log)*	.	1	.	1	.	.	.	.	.	.	.
Observation NCO *(for scissors scope and also for rangefinder)*	.	1	.	.	(1)	.	.	.	.	.	.
Aiming Circle NCO	.	2	.	.	(2)	.	.	.	.	.	.
Wireless Operator in Armoured Vehicle	.	1	.	1	.	.	.	.	.	.	.
Driver for Prime Mover	.	1	.	.	+)	.	.	.	.	.	.
Wireless Operator in Armoured Vehicle	.	.	1	1	.	.	.	.	.	.	.
Courier *(also MG gunner and wireless operator)*	.	.	1	.	1	.	.	.	1	.	.
MC Courier (1 light MC 350cc, 1 MC Sidecar)	.	.	2	2	.	.	.	.	(1)	.	.
Driver for Motor Vehicles	.	.	1	.	(1)	.	.	.	.	.	.
Light Motor Vehicle, cross-country (4 seat)	.	.	.	.	.	.	.	.	.	1	.
Mittl.Beob.Panz.Wg. (Sd.Kfz.251/18)	.	.	.	.	(1)	(1)	.	.	.	.	(1)
Light Wire Section (motorised), x3 *leichter Feldkabeltrupp*											
Telephone Operator *(also Section Leader, 1 also driver for motor vehicle)*	.	.	3	3	.	.	.	.	.	.	.
Light Motor Vehicle, cross-country (4 seat)	.	.	.	.	.	.	.	.	.	1	.
Totals for A. Command Group	**3**	**6**	**8**	**8**	**5 (6)**	**(1)**	.	.	**1 (1)**	**2**	**(1)**

Kriegsstärkenachweisung Nr. 1161 15/09/44 — Panzer-Sturmmörserkompanie "Tiger" (Panz.Stu.Mörs.Kp. "Tiger")	Head count			Weapons				Vehicles			
	Officers	NCOs	Enlisted Personnel	Rifles, Carbines	Pistols (Machine Pistols)	Heavy MGs (Light MGs)	Limbered (Unlimbered) Guns and Launchers	Limbered (Unlimbered) Vehicles	Motorcycles (MC Sidecars)	Motor Transport (Lorries)	Prime Movers (Fully Tracked Prime Movers)
B. 1st Platoon *1.Zug*											
Platoon Leader *(also vehicle commander)*	1	·	·	·	1	·	·	·	·	·	·
Vehicle Commander	·	1	·	·	1	·	·	·	·	·	·
Driver for Armoured Vehicle	·	2	·	·	+)	·	·	·	·	·	·
Wireless Operator *(also machine gunners)*	·	2	·	·	2	·	·	·	·	·	·
Observation NCOs *(also radio operators)*	·	2	·	·	(2)	·	·	·	·	·	·
Loaders	·	·	4	·	4	·	·	·	·	·	·
MC Courier (Light MC 350cc)	·	·	1	1	·	·	·	·	1	·	·
Panzersturmmörser 38cm 'Tiger'	·	·	·	·	(2)	(2)	(2)	·	·	·	2
Totals for B. 1st Platoon	**1**	**7**	**5**	**1**	**8 (4)**	**(2)**	**(2)**	**·**	**1**	**·**	**2**
C. 2nd Platoon same as 1st Platoon	**1**	**7**	**5**	**1**	**8 (4)**	**(2)**	**(2)**	**·**	**1**	**·**	**2**
D. Trains *Troß*											
First Sergeant *(also Leader of the Trains) (also paymaster)*	·	1	·	·	(1)	·	·	·	·	·	·
Ammunition NCO	·	1	·	1	·	·	·	·	·	·	·
Motor Sergeant	·	1	·	·	(1)	·	·	·	·	·	·
Armourer *(also assistant driver)*	·	1	·	·	1	·	·	·	·	·	·
Mess Sergeant *(also 2nd driver of motor vehicle)*	·	1	·	1	·	·	·	·	·	·	·
Medic NCO	·	1	·	·	1	·	·	·	·	·	·
Assistant Armourer *(also assistant driver of 2nd motor vehicle)*	·	·	1	1	·	·	·	·	·	·	·
Clerk *(also Motor vehicle driver)*	·	·	1	1	·	·	·	·	·	·	·
Stretcher Bearer *(on the MC sidecar)*	·	·	1	·	1	·	·	·	(1)	·	·
Field Cook	·	·	1	1	·	·	·	·	·	·	·

Kriegsstärkenachweisung Nr. 1161 15/09/44 — Panzer-Sturmmörserkompanie "Tiger" (Panz.Stu.Mörs.Kp. "Tiger")	Head count			Weapons				Vehicles			
	Officers	NCOs	Enlisted Personnel	Rifles, Carbines	Pistols (Machine Pistols)	Heavy MGs (Light MGs)	Limbered (Unlimbered) Guns and Launchers	Limbered (Unlimbered) Vehicles	Motorcycles (MC Sidecars)	Motor Transport (Lorries)	Prime Movers (Fully Tracked Prime Movers)
Lorry Drivers	·	·	9	9	·	·	·	·	·	·	·
For Ammunition *(also driver of 2nd motor vehicle)*	·	·	2	2	·	·	·	·	·	·	·
Light Motor Vehicle, cross-country (4 seat)	·	·	·	·	·	·	·	·	·	1	·
Lorry, 3t, open top, cross-country											
For small field kitchen and rations	·	·	·	·	·	·	·	·	·	(1)	·
For equipment and baggage	·	·	·	·	·	·	·	·	·	(1)	·
For fuel	·	·	·	·	·	·	·	·	·	(3)	·
For ammunition	·	·	·	·	·	·	·	·	·	(4)	·
Lorry, 3t, open top, cross-country											
For small field kitchen and rations	·	·	·	·	·	·	·	·	·	(1)	·
For equipment and baggage	·	·	·	·	·	·	·	·	·	(1)	·
For fuel	·	·	·	·	·	·	·	·	·	(3)	·
For ammunition	·	·	·	·	·	·	·	·	·	(4)	·
Replacement Crewmembers	·	·	·	·	·	·	·	·	·	·	·
Vehicle Commander*	·	·	·	·	·	·	·	·	·	·	·
Driver for armoured vehicle*	·	·	·	·	·	·	·	·	·	·	·
Wireless operator *(also machine gunner)* *	·	·	·	·	·	·	·	·	·	·	·
Loader*	·	·	·	·	·	·	·	·	·	·	·
Totals for D. Trains	·	**8**	**17**	**16**	**7 (2)**	·	·	·	**(1)**	**1 (9)**	·
E. Vehicle Maintenance Section *Kfz.Instandsetzungsgruppe*											
Armoured Vehicle Mechanic I *(also section leader)* (Feldwebel)	·	1	·	·	(1)	·	·	·	·	·	·
Armoured Vehicle Mechanic I *(also equipment supervisor)*	·	1	·	·	1	·	·	·	·	·	·

	Head count			Weapons				Vehicles			
Kriegsstärkenachweisung Nr. 1161 15/09/44 **Panzer-Sturmmörserkompanie "Tiger" (Panz.Stu.Mörs.Kp. "Tiger")**	Officers	NCOs	Enlisted Personnel	Rifles, Carbines	Pistols (Machine Pistols)	Heavy MGs (Light MGs)	Limbered (Unlimbered) Guns and Launchers	Limbered (Unlimbered) Vehicles	Motorcycles (MC Sidecars)	Motor Transport (Lorries)	Prime Movers (Fully Tracked Prime Movers)
Armoured Wireless Mechanic	·	1	·	·	1	·	·	·	·	·	·
Armoured Vehicle Mechanic I including 1 fabricator (*also armour welder*), *1 also armoured electrician*	·	·	4	4	·	·	·	·	·	·	·
Carry over for E. Vehicle Maintenance Section	·	3	4	4	2 (1)	·	·	·	·	·	·
Armoured Vehicle Mechanic II (*including1 motor vehicle driver, also 2 lorry drivers*)	·	4	·	4	·	·	·	·	·	·	·
Light motor vehicle, cross-country (4 seat)	·	·	·	·	·	·	·	·	·	1	·
Lorry, 3t, open top, cross-country											
Outfitted in accordance with D 623/3	·	·	·	·	·	·	·	·	·	(1)	·
for replacement parts	·	·	·	·	·	·	·	·	·	(1)	·
Totals for Vehicle Maintenance Section	·	**3**	**8**	**8**	**2 (1)**	·	·	·	·	**1 (2)**	·
Summary											
A. Command Group	**3**	**6**	**8**	**8**	**3 (6)**	**(1)**	·	·	**1 (1)**	**2**	**(1)**
B. 1st Platoon	**1**	**7**	**5**	**1**	**8 (4)**	**(2)**	**(2)**	·	**1**	**2**	·
C. 2nd Platoon	**1**	**7**	**5**	**1**	**8 (4)**	**(2)**	**(2)**	·	**1**	**2**	·
D. Trains	·	**8**	**17**	**16**	**7 (2)**	·	·	·	**(1)**	**1 (9)**	·
E. Vehicle Maintenance Section	·	**3**	**8**	**8**	**2 (1)**	·	·	·	·	**1 (2)**	·
Total strength for Panz.Stu.Mörs.Kp. "Tiger"	**5**	**31**	**43**	**34**	**28 (17)**	**(5)**	**(4)**	·	**3 (2)**	**4 (11)**	**4 (1)**

Notes:
1. Four personnel are to be designated as assistant stretcher bearers.
2. The unit will establish a gas detection team consisting of a team leader and three enlisted personnel and a troop decontamination team consisting of a team leader and 6 enlisted personnel (including, if on hand, one medic or stretcher bearer). See H.Dv. 395/1, Annex 11.
3. One Unteroffizier is to be designated as the gas defence NCO.
4. One Unteroffizier or enlisted man is to be designated as an equipment supervisor.
5. Administrative and medial support is determined on a case-by-case basis by the designated superior command.
* Also assistant drivers.

ETO Ordnance Technical Intelligence Report No. 184
Subject: 38cm Rocket Projector on Tiger I Chassis

Observations by: Capt. R. E Howell and Sgt. B.C. Washer, Ord, Tech Intell. Team No. 9

1. General

A 38 cm (15 inches) rocket projector mounted on a modified Tiger I Chassis has been examined in the Ninth U.S. Army area. Although the German nomenclature for this equipment is hot known, it may be the vehicle referred to in German documents as the *"Panzer Sturm Mörser Tiger"*.

The vehicle on which the rocket projector is mounted is a modified version of the Tiger I Chassis, with a rectangular super structure replacing the turret.

The rocket projector is radically different in design and construction from any weapon previously examined. The propellant gases are deflected between the tube and liner by an unusual obturator, and escape through a perforated ring at the muzzle. The splined projectile fired by the projector is approximately five feet long and weighs 726 pounds. An unconfirmed report states that the range of the rocket is 6000 meters (6552 yards). The same source reports that the vehicle has a crew of seven, including a tank commander, a forward observer and five men to operate the vehicle and rocket projector.

2. Chassis

The suspension, power train, engine and hull are those of the Pz.Kpfw Tiger, Model E (Tiger I)

The normal superstructure and turret of the tank have been replaced by a heavy rectangular superstructure of the type used on the "Panzerjäger" self-propelled guns. The superstructure is made of rolled armor plates and is of welded construction with the side plates interlocked with the front and rear plates. A heavy strip of armor is used to reinforce the joint between the front plate and glacis plate on the outside.

A ball-mounted machine gun, MG34, is set into the front plate on the right side.

Armor thickness and angles of the superstructure

	Thickness	Angle to vertical
Front plate	150 mm (5·9 in.)	45°
Projector mantlet (average)	69 mm (2·4 in.)	Rounded
Projector shield (average)	150 mm (5·9 in.)	Rounded
Side plates	80 mm (3·2 in.)	20°
Rear plate	80 mm (3·2 in.)	10°
Top plate	40 mm (1·6 in.)	Horizontal

Dimensions of the superstructure

Front plate

Width at base	126 inches
Width at top	101 inches
Height (base to top)	73 inches

Side plates

Length at base	128 inches
Length at top	86-1/2 inches
Height (base to top)	38-5/8 inches
Overall height from ground (excluding loading crane)	9 ft 3 in.
Overall height (including loading crane)	11 ft 4 in.

Ports and hatches

A rectangular loading hatch, 62 inches long and 19 inches wide, is located in the center rear of the top plate. It is closed by two doors, one forward and one to the rear. The rear door can be opened independently of the forward one and mounts a smoke projector, which has 360° traverse, in the center. The door is spring balanced and is hinged at the rear to open, outward. A loading crane is mounted on the right rear corner of the superstructure:

A circular hatch, 19 inches in diameter, is provided in the superstructure rear plate.

A pistol port, closed by a conical plug, is located at the front of each side plate.

A ventilator for the fighting compartment is mounted at the right front of the top plate.

Vision

The driver is provided with a double periscope mounted in the superstructure front plate. An opening directly above the driver's periscope permits the use of sighting equipment for the rocket projector. This opening can be closed off on the inside by means of an armor plate moved vertically by a rack and pinion arrangement.

A periscope with 360° traverse is mounted at the rear of the superstructure top plate.

3. Rocket Projector

The rocket projector is mounted in the front plate of the superstructure and is offset to the right of center.

The tube

The tube consists of a tubular casting and a spaced, rifled liner. The liner is a steel tube of 1/2 inch wall thickness which is hold in place at the rear by four steel blocks and by a perforated ring at the muzzle end. This ring has thirty-one equally spaced hole's around its face. The holes are 13/16 inches in diameter.

The liner has nine grooves with right hand twist, with one turn in 17·6 calibers. At the extreme rear of the liner, the grooves widen to aid in positioning splines near the base of the projectile. The length of the liner is 74-1/4 inches

A 2-5/16 inch thick horizontal sliding plate; opening from left to right,

functions as a breech block. It is opened and closed by means of a rack and pinion.

A metal obturator fits into a circular recess in the front face of the breech plate. The obturator consists of a thin 'L' shaped outer ring, a heavier 'L' shaped inner ring and a spacer ring. The inside diameter of the assembled obturator is 15-15/32 ;inches, the outside diameter is 17-29/32 inches. The thickness of the obturator, without the spacer, is 0·8 inches.

The side and face of the obturator ring are very thin (0·033 inches) with a 3/8 inch fillet in the inside angle. The side bears against the side of the recess in the breech plate; and the face bears against the rear face of the tube when the breech is closed.

The inner ring, 'B', is 0·15 inches thick on the face, 0·185 inches thick on the side, with a 9/16 inch radius fillet. The side of the inner ring is bored with 80 radial ports ,each 1/2 inch in diameter. Since the 'L' sections are opposite, the inner and outer rings, when assembled, form a chamber opening into the tube through the eighty radial ports.

When the projectile is fired, the propellant gases pass through the ports to the chamber between the inner and outer rings, and force the face of the outer ring against the rear face of the tube, and the side of the outer ring against the recess in the breech plate, thereby obtaining the gas seal. .

The spacer ring, 'C', is positioned at the bottom of the breech plate in order to position the face of the outer ring against the rear face of the tube. Both diameters of the rear face of the spacer ring are bevelled. The thickness of the spacer ring is 0·123 inches. Extra spacer rings were found in the breech spare parts box. The thicknesses of these rings are 0·116, 0·118, 0·125, and two rings of 0·128 inches.

A continuous-pull type firing mechanism is contained in the rear face of the breech plate.

A mantlet forms an integral part of the tube and affords protection far the joint of the tube and mount.

Four projecting lugs are located near the muzzle end of the tube. The lugs at the top and bottom of the tube are square; the lugs on each side of the tube are cylindrical. The purpose of these lugs may be either for attachment of a tube extension or for use when removing or installing the projector.

4. The Mount

The mount consists of a large cast bracket extending through the front plate of the superstructure and welded to the inside. The inside of the bracket conforms to the spherical casting which acts as a cradle. The cradle is mounted to the bracket by horizontal trunnions. The tube is mounted in the spherical cradle by vertical trunnions (Appendix A, Photo 2). In elevation, the cradle and tube elevate together; during traverse, the tube pivots in the cradle - the cradle remaining stationary.

Elevation

Elevation is from 0° to approximately 85°, and provides the only means of regulating the range of the projectile. The elevating mechanism consists of a worm, wormwheel, and an arc and pinion on the left side of the projector and is operated by a handwheel. The elevating arc is bolted to a bracket which projects from the rear of the spherical cradle.

The projector and cradle are extremely well balanced and can be elevated with ease, with or without a projectile in the tube.

Traverse

Traverse is approximately 20° (10° right and left of center). For greater shifts in traverse, it is necessary to move the vehicle.

The traversing mechanism consists of a handwheel, worm and wormwheel, and a pinion and rack. The rack is bolted to the top of the rear bracket, and the other components are bolted to the top of the tube. A traverse indicator is graduated from 0 to 200 mils (10°) to the right and left.

Sight bracket

The sight bracket is bolted to the front plate of the superstructure and is linked to the left trunnion. No range drum or sale was found in the vehicle examined. The azimuth indicator has a micrometer drum which is graduated from 0 to 100 mils and a hundred-mil scale graduated to 2 on each side of the 0 line.

5. Loading Arrangements

An ammunition loading tray is supported by tubular supports which fold into the floor when not in use. The tray is fitted with six rollers to assist in the manual loading of the projectiles.

A hand-operated winch on overhead rails is fitted to the roof of the superstructure. The rails run the entire width of the superstructure. The winch is used to place the projectiles in the storage racks and to carry the projectiles from the racks to the loading tray.

6. Ammunition Stowage

Ammunition racks are provided within the vehicle for twelve projectiles. One projectile may also be carried in the projector tube. Six racks are on each side of the fighting compartment.

7. Operation

The projectiles are placed on the loading tray by means of the winch and are then loaded by hand with the projector at 0° elevation. A plunger, fitted to the inside of the tube at the rear end, drops behind the projectile to prevent it from slipping back from its firing position, 5 inches in front of the breech plate, when the tube is elevated.

As the breech plate is closed, the camming groove of a bracket, screwed to the top of the breech plate, cams a lock over the plunger.

The firing mechanism is then slid upward in a bracket which is screwed to the top of the breech plate and an igniter is inserted into the igniter holder in the breech plate.

The firing mechanism is then slid dawn into the firing position and the projector is layed for elevation and azimuth.

When the lanyard is pulled, the flame from the igniter flashes across a gap to the igniting primer of the projectile. The propellant gases pass through the rear opening between the tube and the liner, through the space between the tube and liner, and out through 31 holes of the perforated ring at the muzzle end.

8. Projectile

The projectile consists of a three-piece steel body: the nose, which contains the explosive charge and makes approximately 60% of the total weight of the projectile; the tail, which contains the propellant charge; and the base plate. The sections are screwed together and are held in place by two locking screws.

1. The nose section is 37-11/16 inches long. It contains a fuze well into which a point-detonating fuze fits. The fuze is constructed of aluminium and is armed by the rotation of the projectile.

2. The tail section is 18-7/16 inches long, and is slotted near the base plate to hold the nine rotating splines winch [sic] fit into the grooves of the tube liner). The splines are 1·57 inches long, 0·39 inches wide, and 0·39 inches high.

3. The base plate has a thickness of 15/16 inches and contains thirty-two venturis which are bored at an angle to assist in the rotation of the rocket projectile. The venturis are 0·79 inches in diameter.

Color and markings

The projectile is painted dark green and has a 13/16 inch wide white band painted around it at the center of gravity.

On each side of the nose section the following words were stencilled in white:

Achtung !	*Warning !*
Feuchtigkeitsempfindlich	*Sensitive to humidity*
vor Regen und Nässe	*Protect from rain*
zu schützen	*and dampness*

A more detailed report on the projectile is being prepared.

9. Data

Modified Tiger I Chassis

1. Overall length	248 in.
2. Overall width	147 in.
3. Overall height (with loading crane)	136 in.
4. Overall height (without loading crane)	111 in.
5. Width of track	28-1/2 in.
6. Pitch of track	5-1/8 in.
7. Track links	95
8. Pitch diameter	32 in.
9. Drive sprocket teeth	20

Rocket projector

1. Caliber	38 cm (15 in.)
2. Length of tube (overall)	81-1/8 in.
3. Length of liner	74-1/4 in.
4. Thickness of liner	1/2 in.
5. Number of grooves	9
6. Depth of grooves	0·2 in.
7. Width of grooves	0·4 in.
8. Width of grooves at rear	1·06 in.
9. Twist of grooves	Right hand, one turn in 17·6 calibers
10. Space between liner and tube	1-1/2 in.
11. Breech plate	Horizontal sliding
12. Firing mechanism	Continuous-pull
13. Traverse	20°
14. Elevation (approx.)	0° to 85°
15. Markings on breech plate	R5 bwo

Projectile

1. Overall length	60 in.
2. Weight	726 lbs.

3. Length of nose section	37-11/16 in.
4. Length of tail section	18-7/16 in.
5. Thickness of base plate	15/16 in.
6. Markings on nose section	13 E (near fuze well)
	346.5 (center of section)
	III 1.9.4.4. (near base of section)
7. Markings on tail section	569 (near top of section)
8. Fuze	Point-detonating.

ETO Ordnance Technical Intelligence Report No. 184a Subject: 38cm Rocket Projector Supplemental. (Raketen Werfer) on Modified Tiger I Chassis Supplemental 24 March 1945

1. General

A range table for the 38 cm Rocket Projector mounted on the modified Tiger I Chassis has been found by 21st Army Group. The range table was published by Rheinmetall Borsig, the manufacturers of the rocket projector, and is dated August, 1944.

The information contained therein supplements that given in ETO Ordnance Technical Intelligence Reports Nos. 184 and 192.

2. Nomenclature of Rocket Projector

The 38 cm Rocket Projector is identified in the range table as 'Raketen Werfer 61' (RW 61).

3. Additional Ammunition Data

The range table also contains the following information:

a. Types of ammunition:
   H.E. (R. Sprenggranat 4581)
   H.C. (R. Holladungsgranat 4592)

b. Range table weight of projectile: 761 lb. (Large variations are encountered in individual projectiles. Weight zones are marked to nearest 5 kg. (12 lbs.))

c. Maximum range at 15°c (59°F) is 6179 yards.

ETO Ordnance Technical Intelligence Report No. 192 Subject: 38 cm. R. Sprgr. 4581 Rocket Projectile for (RW 61) Projector 23 March 1945

Observations by: Maj. H.L. Kersch, Enemy Equipment Intell. Sec., Ord. Service, Hq. Com Z, ETOUSA.

1. General

The 38 cm. R. Sprgr. 4581 rocket projectile is fired from the new type rocket projector Raketenwerfer 61 (RW 61) mounted on a Tiger I chassis. Information on this projector is contained in ETO Ord. Tech. Intell. Report No. 184. Although much larger in diameter, this new rocket is similar in design and appearance to the 21 cm. rocket projectile. Wgr. 42 Spr. Mit Hbgr. Z 35 K. The radical departure from the standard spin-stabilized rocket design is in the addition of insert splines at the after end of the motor body. Those splines, fitting into the rifling of the liner, aid in imparting an initial spin to the projectile. Range table weight of the projectile is listed at 761 lbs. with variations in steps of 12 lbs. each to be expected. The maximum range of the rocket is listed at 6180 yds. When the rocket-propellant charge is at +14°c. The range table indicates that the rocket is sensitive to temperature. Excluding the motor body, the projectile has a charge weight ratio of 73% or an overall charge weight ratio of 35·5%.

2. Projectile

General

The rocket projectile consists of three main assemblies, namely the high explosive body, the motor body, end the nozzle assembly. . Following are the dimensions of the rocket:

Height of rocket	761 lbs.(±12 lbs. variations)
Overall length (not including fuze)	56·68 ins. (144 cm.)
Diameter of bourrelet	14·94 ins. (38 cm.)
Length of H.E. body	37·19 ins.(94·5 cm.)
Length of motor body	18·54 ins.(47·1 cm.)
Thickness of H.E. body	0·30 in.(0·762 cm.)
Thickness of motor body	0·53 in. (1·346 cm.)

High Explosive Body

The H.E. body (forward section) is of two-piece welded construction and is threaded internally at its after end to receive the motor body. The booster pocket and fuze adapter assembly is welded in position at the nose of the H.E. body. A bourrelet is located just in rear of the welded junction of the ogive and the cylindrical section.

Motor Body

The motor body (rear section) is threaded externally to screw into the H.E. body and threaded internally to receive) the nozzle assembly. Both the H.E. body and nozzle assembly are secured by means of two diametrically opposed set screws. Nine grooves for the splines are machined into the base of the periphery of the motor body.

Spacer

A spacer ring, located in the forward end of the motor body, has a dual purpose: (i) to hold the forward ignition charge in place and to position the rocket propellant charges, which vary in length; and (ii) to allow space for proper ignition of the various elements of the rocket propellant.

Splines

Several of the nine splines fitted in the grooves in the periphery of the motor body are made in two pieces. There is no apparent reason for this two-piece construction. These splines are held in position by tho shoulder on the nozzle assembly plate.

3. Nozzle Assembly

There are thirty-two (32) venturi holes in the nozzle plate, details of which are shown in Nozzle Detail, Drawing I, Appendix 'B'. These venturi holes are set at an angle of 14° to the axis of the rocket. This angle, in addition to the splines, gives the rocket a clockwise rotation in flight as well as providing the forward impulse.

In the center of the nozzle plate there is a threaded hole to receive the igniter primer for the rocket propellant.

A rear spacer ring, welded to the nozzle plate, aids in the positioning of the outer row of propellant charges. The remaining elements of the propellant charges are positioned by the base plate of the nozzle assembly.

Explosive

The high explosive body is filled with 270 lbs. of the German explosive charge No. 13A, which is amatol 50/50 poured. The H.E. charge is initiated by the sub-booster cwg. Np. 10, the same type as that used in German artillery projectiles, and a booster of six (6) P.E.T.N. pellets 1-3/4 ins. in diameter by 11/16 inch thick. Four (4) P.E.T.N. pellets are cast in the explosive aft of the booster pocket as a secondary booster.

Propellant Charge

The propellant charge, total weight 88-½ lbs., located in the motor body, consists of the following components:

1 ea. Center Stick	
Length	16 ins.
Diameter	4-1/16 ins.

| Diameter of hole | 1-3/8 ins. |
| Weight | 10-½ lbs. |

Cemented to the central stick are four ribs which act as spacers, dimensions of which are as follows:

1 ea. Intermediate Stick

Length	15-5/16 ins.
Diameter	7-¼ ins.
Diameter of hole	4-3/4 ins.
Weight	20 lbs.

10 ea. Outer Sticks

Length	12-15/16 ins.
Diameter	3-3/16 ins.
Diameter of hole	7/16 inch.
Weight per stick	5·6 lbs.
Total Weight	56 lbs.

A thin asbestos gasket covers the venturi holes to protect the propellant charge from moisture.

Propellant Ignition System

The rocket propellant ignition system is composed of the following elements:

1. Igniter Primer
This primer, located in the nozzle assembly plate, is initiated by a flash from the firing mechanism in the breech of the projector.

2. Relay
The relay from the igniter primer to the ignition charge is made up of three (3) bags of powder located in the hole of the center stick of the propellent charge.

3. Ignition Charge
This charge, fastened to the forward spacer ring, weighs approximate ½ lb. and is used to ignite the rocket propellant.

Color and Markings

The projectile is painted O.D. with a 0.8 inch wide white band around the body at the center of gravity. Stencilled on the ogive is the following warning: *"ACHTUNG: FEUCHTIGKEITSEMPFINDLICH. VOR REGEN UND WASSER ZU SCHUTZEN"* which, translated, means "Warning: Sensitive to damp. Protect from rain and water." The German explosive code number 13A is stencilled on the ogive in black.

Fuze

A point detonating (percussion) nose fuze, designation unknown, is fitted in the forward fuze pocket. The fuze body adapter and retaining collar are made of steel, while the fuze body itself is of aluminium. During transit the fuze may be set in safe position by rotation of the fuze body. A choice of either instantaneous or short delay setting can be made prior to firing by rotation of the fuze body to the desired position. After being armed by rotation, the striker is held away from the primer cap by means of a creep spring. This fuze will be covered separately in a later report.

1. *Unknown Author: Waffen Revue, volume 93, page 3.*
2. *Kosinski: Die Ketten- und Halbkettenfahrzeuge, page 74 f.*
3. *Albert Speer in Waffen Revue, Volume 93, page 4.*
4. *Unknown Author: Waffen Revue, volume 93, page 5.*
5. *Schneider: Elefant-Jagdtiger-Sturmtiger, page 33.*
6. *General Staff of the Army, Org. Abt (IIIa), No. 1183/44 g Kdos Chefs., dated 26 January 1944.*
7. *Unknown Author: Waffen Revue, volume 93, page 6.*
8. *Albert Speer in Waffen Revue, Volume 93, page 6 f.*
9. *Unknown Author: Waffen Revue, volume 93, page 8 f.*
10. *Unknown Author: Waffen Revue, volume 35, page 5522.*
11. *Kosinski: Die Ketten- und Halbkettenfahrzeuge, page 73.*
12. *Unknown Author: Waffen Revue, volume 93, page 9.*
13. *Spielberger: Der Panzerkampfwagen Tiger und seine Abarten, page 172.*
14. *BAMA, RH 10/350, Inspector General Armoured Forces, page 156.*
15. *Spielberger: Der Panzerkampfwagen Tiger und seine Abarten, page 172.*
16. *According to Chamberlain (Encyclopaedia of German Tanks, page 138) 6·28m.*
17. *Chamberlain (Encyclopaedia of German Tanks, page 138) provides more detailed information. Fighting Compartment: Front 150mm at 45°; sides 80mm at 30° to 0°; rear 80mm at 0°; top 40 to 25mm at 90°. Hull: Front 100mm at 25°; sides 60mm at 0°; rear 80mm at 9°; bottom 25mm at 90°.*
18. *According to Fortin (in Steelmasters, No. 2/95, page 30 f.), all assault mortars had steel road wheels except for the prototype, which had roadwheels with rubber tyres.*
19. *According to Kleine/Kühn (Tiger, page 316.): 28 km/h was the maximum speed. According to Chamberlain (Encyclopaedia of German Tanks, page 138): 40km/h.*
20. *According to Chamberlain (Encyclopaedia of German Tanks, page 138): 120km.*
21. *In a book published in 1966 (Nowarra: The Tiger Tanks), it was surmised that the rocket was loaded from the outside.*
22. *According to Forty (German Tanks of WW II, page 137) up to 85°.*
23. *For ammunition basic issue, there are a variety of contradictory statements. According to Kleine/Kühn (Tiger, page 317.), an assault mortar could carry 13 rounds, one of which was in the gun barrel. According to Hauptmann Adams (in Mues: Der große Kessel, page 591.) it could carry only 7 rounds, one of which was on the loading tray.*
24. *Kosinski: Die Ketten- und Halbkettenfahrzeuge, page 73.*
25. *Schneider: Elefant-Jagdtiger-Sturmtiger, page 34 f.*
26. *Next to the rocket explosive shell 4581, a rocket hollow charge 4592 was also tested. (No author given: Waffen Revue, volume 93, page 12). It appears that only the rocket explosive shell 4581 was used in combat. (Schneider: Elefant-Jagdtiger-Sturmtiger, page 37.)*
27. *According to Kleine/Kühn (Tiger, page 316.) 1490mm and according to Schneider (Elefant-Jagdtiger-Sturmtiger, page 37.) only 890mm.*
28. *According to Kleine/Kühn (Tiger, page 316.) 330kg.*
29. *According to Hahn (Waffen und Geheimwaffen, page 59) 5650m.*
30. *Kosinski: Die Ketten- und Halbkettenfahrzeuge, page 75 f.*
31. *US Army ETO Ordnance Technical Report No. 184, dated 18 March 1945, refers to it as three parts: nose section, tail section and base plate. US Army ETO Ordnance Technical Report No. 192, dated 23 March 1945, refers to these as the high-explosive body, motor body and the nozzle assembly. Hereafter, respectively ETO 184 and ETO 192.*
32. *Kilopond (abbreviated to kp) refers to the 'weight' an object with a mass of one kilogram has at sea level. For nonscientific and practical purposes, it is the same as a kilogram. ETO 192 states the explosive charge weighed 270 pounds.*
33. *ETO 192 refers to this as 88·5 pounds of propellant.*
34. *According to Waffen Revue (Volume 93, page 14 f.) there were variations with 20, 30 or even no holes.*
35. *Schneider: Elefant-Jagdtiger-Sturmtiger, page 37.*
36. *Hahn: Waffen und Geheimwaffen, page 59.*
37. *Unknown Author: Waffen Revue, volume 35, page 5529.*
38. *Schußtafel is literally a 'shooting table' (range chart). Heeres-Dienstvorschrift is an Army Manual. It will hereafter appear in its normally abbreviated form: HDv.*
39. *This abbreviation appeared in the original source, and the author has been unable to determine the exact meaning, although it is most likely an Army directorate.*

40. Unknown Author: Waffen Revue, volume 35, page 5540.

41. Unknown Author: Waffen Revue, volume 35, page 5540.

42. Enclosure to the daily logs of AOK 9, Ia, Nr. 4228/44 (secret), dated 20 August 1944: Order of battle for assault troops in Warsaw.

43. K.St.N. 1161, dated 15 September 1944; Pallud: Battle of the Bulge, page 46 f; Jentz: Die deutsche Panzertruppe, volume 2, page 174.

44. BAMA, RH 10/109, Inspector General of Armoured Force: Overview of forces in the process of activation or reorganization, dated 1 September 1944; Jentz: Die deutsche Panzertruppe, volume 2, page 174.

45. NARA, T311, R228, Daily logs, Heeresgruppe Mitte, 12 August 1944.

46. NARA, T311, R228, Daily logs, Heeresgruppe Mitte, 12 August 1944.

47. Jentz: Die deutsche Panzertruppe, Volume 2, page 174.

48. www.achtungpanzer.com/sttig.htm

49. NARA, T312, Roll 348, Daily logs, No. 11 AOK 9, (11 July - 15 November 1944), frames 7922847, 7922862 and 7922867.

50. NARA, T312, Roll 348, Daily logs No. 11 AOK 9, (11 July - 15 November 1944), frames 7922874, 7922876, 7922881, 7922883 and 7922888.

51. www.achtungpanzer.com/sttig.htm

52. peer in Waffen Revue, Volume 93, page 9.

53. NARA, T312, Roll 348, Daily logs, No. 11 AOK 9, (11 July - 15 November 1944), frame 7922838.

54. Schramm: KTB des O.K.W. 1944-1945, volume I, page 381; NARA, T311, R30, Railroad movements OB West, 15 November 1944, F. 7037577.

55. Message Generalstab des Heeres, g.Kdos 10223/24, dated 24 August 1944.

56. BAMA, RH 10/349, Inspector General of Armoured Forces, page 316; BAMA, RH 10/350, Inspector General of Armoured Forces, page 156.

57. Message Heeresgruppe Mitte, Ia Nr. 11 659/44 g.Kdos, dated 25 August 1944.

58. NARA, T312, Roll 348, Daily logs, No. 11 AOK 9, (11 July - 15 November 1944), frames 7922831, 7922809, 7922805, 7922800, 7922796, 7922790, 7922782, 7922776, 7922771.

59. Jentz: Die deutsche Panzertruppe, Volume 2, pages 174 and 191.

60. www.achtungpanzer.com/sttig.htm

61. Ritgen: Die Geschichte der Panzer-Lehr-Division im Westen 1944/45, page 193.

62. BAMA, RH 10/109, Inspector General of Armoured Forces, page 93.

63. A.O.K 9: After-action report to Heeresgruppe Mitte, dated 2 September 1944.

64. NARA, T312, Roll 348, Daily logs, No. 11 AOK 9, (11 July - 15 November 1944), frames 7922610 ff.

65. German Docs in Russia: Findbuch 12485, Akte 11. Unterlagen der Ia-Abteilung der Heeres-Artillerieabteilung 849: KTB Nr. 8 der Heeres-Artillerieabteilung 849 vom 1.7.-31.12.1944 zum Einsatz an der Ostfront u.a., page 161.

66. Żoliborz is one of the city districts of Warsaw. It is located north of the city centre on the left bank of the Vistula and had been surrounded by German forces.

67. German Docs in Russia: Findbuch 12475, Akte 211: Unterlagen der Ia-Abteilung des Generalkommandos des XXXXVI. Panzerkorps: KTB Nr. 13 des XXXXVI. Panzerkorps, 1.7.-30.9.1944., page 185.

68. German Docs in Russia: Findbuch 12478, Akte 122. Unterlagen der Ia-Abteilung des Panzerartillerieregiments 19 der 19.Panzerdivision: Anlagen zum KTB, 30.8.1944-11.3.1945 - Divisionsbefehle, Feindlagekarten u.a., page 28.

69. German Docs in Russia: Findbuch 12475, Akte 213: Unterlagen der Ia-Abteilung des Generalkommandos des XXXXVI. Panzerkorps: Anlagen zum KTB Nr. 13 des XXXXVI. Panzerkorps, 1.7.-30.9.1944, page 92.

70. German Docs in Russia: Findbuch 12482, Akte 287: Unterlagen der Ia-Abteilung des Panzer-Artillerieregiments 19: KTB des Panzer-Artillerieregiments 19, 30.8.-11.3.1945, page 15 f.

71. BAMA, RH 10/108, Inspector General of Armoured Forces.

72. Speer in Waffen Revue, Volume 93, page 9 f.

73. No author: Waffen Revue, Vol. 35, page 5540, NARA T78, R 169, Report dated 1 November 1944, Frame 6106514, NARA T78, Roll 169, Report dated 1 December 1944, Frame 6106698.

74. BAMA, RH 10/108, Inspector General of Armoured Forces, Report September 1944; Jentz: Die deutsche Panzertruppe, volume 2, page 174.

75. BAMA, RH 10/104, Inspector General of Armoured Forces, page 2.

76. Kosinski: Die Ketten- und Halbkettenfahrzeuge, page 75.

77. No author: Waffen Revue, Vol. 35, page 5540, NARA T78, R 169, Report dated 1 November 1944, Frame 6106514, NARA T78, Roll 169, Report dated 1 December 1944, Frame 6106698.

78. BAMA, RH 10/349, Inspector General of Armoured Forces, page 316; BAMA, RH 10/350, Inspector General of Armoured Forces, page 156.

79. BAMA, H16/129, pages 9-10.

80. NARA, 12th Army Group, G-2 Periodic Report, No. 347, 18 May 1945.

81. NARA, T312, Roll 348, Daily logs, No. 11 AOK 9, (11 July - 15 November 1944), frames 7922459, 7922462, 7922467, 7922472, 7922479, 7922482, 7922489, 7922492, 7922495, 7922512, 7922508 and 7922504.

82. NARA, T312, Roll 348, Daily logs, No. 11 AOK 9, (11 July - 15 November

1944), frames 7922573 ff.

83. BAMA, RH 10/349, Inspector General of Armoured Forces, page 316; BAMA, RH 10/350, Inspector General of Armoured Forces, page 156.

84. German Docs in Russia: Findbuch 12475, Akte 209: Unterlagen der Ia-Abteilung des Generalkommandos des XXXXVI. Panzerkorps: KTB Nr. 13 des XXXXVI. Panzerkorps, 1.7.-30.9.1944, page 79.

85. Jentz: Die deutsche Panzertruppe, Volume 2, page 174.

86. BAMA, RH 10/349, Inspector General of Armoured Forces, page 316.

87. BAMA, RH 10/349, Inspector General of Armoured Forces, page 316; BAMA, RH 10/350, Inspector General of Armoured Forces, page 156.

88. BAMA, RH 10/121, Inspector General of Armoured Forces, page 187.

89. BAMA, RH 10/109, Inspector General of Armoured Forces, October 1944 reports; BAMA, RH 10/349, Inspector General of Armoured Forces, pages 134 and 316.

90. BAMA, RH 10/109, Inspector General of Armoured Forces, page 63.

91. NARA, T312, Roll 348, Transport officer AOK 9, frames 7922411 ff.

92. BAMA, RH 10/109, Inspector General of Armoured Forces, October 1944 reports; Jentz: Die deutsche Panzertruppe, volume 2, page 174.

93. BAMA, WF 03-4694, OB West, Preparation for the Ardennes Offensive, Vol. V.

94. BAMA, OB West to General der Panzertruppen West, 2 November 1944.

95. BAMA, RH 10/107, Inspector General of Armoured Forces, page 5.

96. BAMA, WF 03-4694, OB West: Preparation for the Ardennes Offensive, Vol. VI.

97. BAMA, H16/129, pages 9-10.

98. NARA, T312, R1571, AOK 7 arrivals, frame 1423.

99. NARA, T311, Roll 30, OB West, State of rail movement, frame 7037577.

100. BAMA, RH 19 IX/13, Morning-, Noon-, Evening reports H.Gr. B, dated 15 November 1944.

101. NARA, FUSA, IPW-Report, 22 November 1944.

102. Kleine/Kühn: Tiger, page 316; Schneider: Elefant-Jagdtiger-Sturmtiger, page 45.

103. BAMA, MS A 988, Gen. Köchling: Order of Battle, LXXXI. A.K., September 1944 to April 1945, page 8.

104. BAMA, MS A 992, Gen. Köchling: Questionnaire regarding the fighting in the area of Aachen from September to November 1944, page 6.

105. BAMA, RH 19 IV-84, Attachments to the daily logs of OB West, page 35.

106. BAMA, WF 03-4694, OB West: Preparation for the Ardennes Offensive, Vol. V.

107. BAMA, WF 03-4694, Daily logs, OB West, Volume 6, page 42.

108. BAMA, H16/129, pages 9-10.

109. BAMA, RH 10/107, Inspector General of Armoured Forces, page 5.

110. NARA, T311, Roll R224, Order of battle, Heeresgruppe Mitte, dated 6 November 1944.

111. NARA, T312, Roll 348, Daily logs, No. 11 AOK 9, (11 July - 15 November 1944), frames 7922455-7923148.

112. NARA, T312, Roll 348, Daily logs, No. 11 AOK 9, (11 July - 15 November 1944), frame 7923144

113. German Docs in Russia: Findbuch 12475, Akte 209: Unterlagen der Ia-Abteilung des Generalkommandos des XXXXVI. Panzerkorps: KTB Nr. 13 des XXXXVI. Panzerkorps, 1.7.-30.9.1944, pages 74 ff.

114. No author: Waffen Revue, Vol. 35, page 5540, NARA T78, R 169, Report dated 1 November 1944, Frame 6106514, NARA T78, Roll 169, Report dated 1 December 1944, Frame 6106698.

115. BAMA, RH 10/109, Inspector General of Armoured Forces, Report November 1944.

116. BAMA, RH 10/107, Inspector General of Armoured Forces, page 5.

117. BAMA, RH 10/109, Inspector General of Armoured Forces, page 63.

118. BAMA, H16/129, pages 9-10.

119. BAMA, RH 10/349, Inspector General of Armoured Forces, page 316; BAMA, RH 10/350, Inspector General of Armoured Forces, page 156.

120. BAMA, RH 10/109, Inspector General of Armoured Forces, page 48.

121. BAMA, WF 03-4694, OB West, Preparation for the Ardennes Offensive, Vol. III.

122. BAMA, RH 10/107, Inspector General of Armoured Forces, page 7.

123. The message confused Sturm-Mörser-Kompanie 1000 with its sister unit.

124. NARA, T-311, Roll 18, OB West - Ardennes Offensive.

125. NARA, T-311, Roll 20, OB West - Transportation reports, frame 092.

126. BAMA, RH 10/107, Inspector General of Armoured Forces, page 7.

127. Pallud: Battle of the Bulge, page 39; Jentz: Germany´s Tiger Tanks, page 112; Parker: Battle of the Bulge, page 240.

128. Jung: Die Ardennenoffensive 1944/45, page 336.

129. BAMA, WF 03-4694, OB West, Preparation for the Ardennes Offensive, Vol. IV.

130. NARA, T-311, Roll 18, OB West - Ardennes Offensive.

131. Pallud: Battle of the Bulge, page 39; Jentz: Germany´s Tiger Tanks, page 112; Parker: Battle of the Bulge, page 240.

132. Pallud: Battle of the Bulge, page 39.

133. Parker: Battle of the Bulge, page 240.

134. NARA, T314, Roll 1594, Daily logs, LXXXI. A.K., frame 00763.

135. Situation maps of OB West, Lage Frankreich, 22 - 31 December 1944.
136. BAMA, RH 10/107, Inspector General of Armoured Forces, page 9.
137. BAMA, RH 10/352, Inspector General of Armoured Forces, page 20.
138. BAMA, RH 10/352, Inspector General of Armoured Forces, pages 20 and 54.
139. BAMA, RH 10/109, Inspector General of Armoured Forces, page 48.
140. BAMA, RH 10/349, Inspector General of Armoured Forces, page 316; BAMA, RH 10/350, Inspector General of Armoured Forces, page 156.
141. BAMA, RH 10/350, Inspector General of Armoured Forces, page 156.
142. BAMA, RH 10/350, Inspector General of Armoured Forces, page 156.
143. BAMA, RH 10/349, Inspector General of Armoured Forces, page 316; BAMA, RH 10/350, Inspector General of Armoured Forces, page 156.
144. No author: Waffen Revue, Vol. 35, page 5540, NARA T78, R 169, Report dated 1 November 1944, Frame 6106514, NARA T78, Roll 169, Report dated 1 December 1944, Frame 6106698.
145. BAMA, RH 10/107, Inspector General of Armoured Forces, page 7.
146. BAMA, RH 10/107, Inspector General of Armoured Forces, page 9.
147. Fortin in Steelmasters Nr. 2/95, page 31.
148. Kleine/Kühn: Tiger - Geschichte einer legendären Waffe, page 316.
149. Situation maps, OB West, Lage Frankreich, January 1944.
150. NARA, 12th Army Group, G2 Periodic Report, No. 347, 18 May 1945.
151. Heinz Matten in Kleine/Kühn: Tiger - Geschichte einer legendären Waffe, page 316.
152. Doll in Kriegszeit und Kriegsende im Drolshagener Land, page 84.
153. Reference is made to the G2 Report of the US 30th Infantry Division, dated 26 February 1945, which mentioned another report from the end of January. The initial report is obviously lost.
154. NARA, G2 Report US 30th Infantry Division, Annex to No. 255, 26 February 1945.
155. BAMA, RH 10/350, Inspector General of Armoured Forces, page 125.
156. BAMA, RH 10/350, Inspector General of Armoured Forces, page 125.
157. BAMA, RH 10/107, Inspector General of Armoured Forces, page 11.
158. BAMA, RH 10/109, Inspector General of Armoured Forces, page 36.
159. BAMA, RH 10/128, Inspector General of Armoured Forces, page 17.
160. GenStdH Org Abt Nr I/14102/44 G V 14.11.44.: „Betr.: Sturm-Mörser-Kompanien", dated 17 January 1945.
161. O.K.H. GenStdH Org Abt Nr. I/20580/45 geh.: "Betr.; Sturm-Mörser-Einheiten.", dated 23 January 1945.
162. NARA, Record Group 242, Attachment 5 to Nr. 630/45 g.Kdos, Inspector General of Armoured Forces, Tank situation West, dated 5 February 1945.
163. This source stated Sturm-Mörser-Batterie 1002.
164. Ob d E/AHA/Stab Ia (2) Nr.11303/45 geh.: "Betr.: Aufstockung Sturm-Mrs. Battr. 1000 auf 6 Geschütze.", dated 20 February 1945.
165. The POW could not make any statements regarding the unit's designation, but the reference to Warsaw is a clear link to Sturm-Mörser-Kompanie 1000.
166. NARA, G2 Report US 1st Infantry Division, No. 252, 28 February 1945.
167. Note Heinz-Adolf Ehser, dated 30 December 2000.
168. US 30th Infantry Division: 5 Stars to Victory, 26 February 1945
169. R. Hewitt: Work horse of the Western Front, page 227.
170. NARA, G2 Report, US 30th Infantry Division, Annex to No. 255, 26 February 1945.
171. http://www.indianamilitary.org/30TH/Articles/Scrapbook-06-1945/02.htm
172. Kölner Stadt-Anzeiger: Verzweifelter Kampf um den Sturmtiger, dated 29 April 2006.
173. NARA, G2 Report, U 30th Infantry Division, Annex to No. 255, 26 February 1945.
174. NARA, G2 Report, US 30th Infantry Division, No. 257, 27 February 1945.
175. NARA, G2 Report, US VII Corps, No. 64, 5 March 1945.
176. NARA, G2 Report, US 7th Armored Division, 8 March 1945.
177. NARA, G2 Report, US 1st Army, No. 262, 27 February 1945.
178. Heinz Matten in Kleine/Kühn: Tiger - Geschichte einer legendären Waffe, page 316.
179. NARA, FUSA IPW-Report, 15 April 1945.
180. Percy E. Schramm: Kriegstagebuch des O.K.W. 1944-1945, Teilband. II, page 1384.
181. Schneider: Elefant-Jagdtiger-Sturmtiger, page 45.
182. This source stated Sturm-Mörser-Batterie 1002.
183. Daily logs, Pz.Jg.Abt. 33, page 42.
184. Daily logs, Pz.Jg.Abt. 33, page 42.
185. Rust: 15.Panzer-Grenadier-Division, pages 72 and 78.
186. Daily logs, Pz.Jg.Abt. 33, page 43.
187. Whitaker: Endkampf am Rhein, page 282.
188. BAMA, RH 39/706, Combat report for Kampfgruppe Tebbe for 18 February to 10 March 1945.
189. Heinz Matten in Kleine/Kühn: Tiger - Geschichte einer legendären Waffe, page 316.
190. NARA, FUSA IPW-Report, 15 April 1945.
191. NARA, US 12th Army Group, G2 Periodic Report, No. 347, 18 May 1945.
192. BAMA, RH 26-353/4, 353. Inf.Div., page 95 f.

193. BAMA, RH 26-353/4, 353. Inf.Div., page 69.
194. Heinz Matten in Kleine/Kühn: Tiger - Geschichte einer legendären Waffe, page 316.
195. NARA, US 12th Army Group, G2 Periodic Report, No. 347, 18 May 1945.
196. NARA, FUSA IPW-Report, 15 April 1945.
197. Doll in Kriegszeit und Kriegsende im Drolshagener Land, page 81.
198. Guderian: Das letzte Jahr im Westen, page 452.
199. BAMA, RH 39/706, Combat report of Kampfgruppe Tebbe from 18 February to 10 March 1945.
200. NARA, US XIX Corps, G2 Periodic Report No. 294, 5 April 1945.
201. Schneider: Elefant-Jagdtiger-Sturmtiger, page 45.
202. Adam in Mues: Der große Kessel, page 591 f.
203. Tebbe in Mues: Der große Kessel, page 591.
204. Adam in Mues: Der große Kessel, page 591 f.
205. Guderian in Mues: Der große Kessel, page 592.
206. Mues: Der große Kessel, page 592.
207. Guderian: 116.Panzer-Division, page 484.
208. BAMA, RW 4/ v. 636, Tank Situation West, Attachment May 1945.
209. Doll in Kriegszeit und Kriegsende im Drolshagener Land, page 81 f.
210. Note that all images of the Sturm-Mörser in Hützemert show that the right track came off.
211. Doll in Kriegszeit und Kriegsende im Drolshagener Land, page 82. ff
212. Doll in Kriegszeit und Kriegsende im Drolshagener Land, page 85 ff.
213. Doll in Kriegszeit und Kriegsende im Drolshagener Land, page 90.
214. Schneider: Elefant-Jagdtiger-Sturmtiger, page 45.
215. NARA, FUSA IPW-Report, 12 April 1945.
216. http://www.tiif.de/thread.php?threadid=691&sid=0f644cd40ad1325efc aa310579314522
217. NARA, FUSA IPW-Report, 13 April 1945.
218. NARA, US 1st Army, IPW-Report, 17 April 1945.
219. NARA, US 9th Army, IPW-Report, No. 225, 19 April 1945.
220. NARA, US 9th Army, IPW-Report, No. 228, 20 April 1945.
221. NARA, US 9th Army, IPW-Report, No. 232, 24 April 1945.
222. NARA, US 12th Army Group, G2 Periodic Report, No. 347, 18 May 1945.
223. NARA, FUSA IPW-Report, 15 April 1945.
224. Doll in Kriegszeit und Kriegsende im Drolshagener Land, page 81.
225. NARA, US 12th Army Group, G2 Periodic Report, No. 347, 18 May 1945.
226. NARA, US 1st Army, IPW-Report, 17 April 1945.
227. Guderian: 116.Panzer-Division, page 488.
228. BAMA, RW 4/ v. 636, Tank situation West, Attachment May 1945.
229. Alwin Hanschmidt: Beiträge zu ihrer Geschichte: 700 Jahre Stadt Rietberg, 1289-1989
230. NARA, G2 Report, US Infantry Division, No. 293, 06.04.1945.
231. Inspector General of Armoured Forces, 07.04.1945; http://www.feldgrau.net/forum/viewtopic.php?f=24&t=23819.
232. Guderian: 116.Panzer-Division, page 530.
233. H.E. Gorges: Der Sturmpanzer von Brumby, Das Calenser Blatt 09/2012
234. H. Menzel: Neue Erkenntnisse zum Sturmtiger bei Ebendorf, http://f15919.nexusboard.de/t1113f60-Ausgabe-Oktober-2.html
235. BAMA, RH 10/349, Inspector General of Armoured Forces, page 316; BAMA, RH 10/350, Inspector General of Armoured Forces, page 156.
236. Guderian in Mues: Der große Kessel, page 592.
237. Schneider: Elefant-Jagdtiger-Sturmtiger, page 45.
238. Rose in Mues: Der große Kessel, page 592; www.737thtankbattalion.org/menden.htm.
239. http://www.tiif.de/thread.php?threadid=691&sid=0f644cd40ad1325efc aa310579314522
240. NARA, US III Corps, G2 Periodic Report No. 126, 16 April 1945.
241. Werner Bergmann: Das Kriegsende 1945 im Landkreis Wunsiedel im Fichtelgebirge, page 14.
242. Messages from Josef Wunderlich, chairman of the local history interest group Mittleres Vilstal, on 22.02.2021. His father, Julius Wunderlich, was one of a delegation of civilians who handed Frontenhausen over to the Americans on 2 May 1945, who shelled the village with artillery because no white flag had been hoisted on the church tower.
243. NARA T78, R987, Fr. 1848, Personnel file for Franz Kodar.

Index